Quanta in Distress

Sadri Hassani

Quanta in Distress

How New Age Gurus Kidnapped Quantum Physics

Sadri Hassani
Department of Physics
University of Illinois at Urbana-Champaign
Urbana, IL, USA

ISBN 978-3-031-65258-5   ISBN 978-3-031-65259-2 (eBook)
https://doi.org/10.1007/978-3-031-65259-2

© The Editor(s) (if applicable) and The Author(s), under exclusive license to Springer Nature Switzerland AG 2024

This work is subject to copyright. All rights are solely and exclusively licensed by the Publisher, whether the whole or part of the material is concerned, specifically the rights of translation, reprinting, reuse of illustrations, recitation, broadcasting, reproduction on microfilms or in any other physical way, and transmission or information storage and retrieval, electronic adaptation, computer software, or by similar or dissimilar methodology now known or hereafter developed.

The use of general descriptive names, registered names, trademarks, service marks, etc. in this publication does not imply, even in the absence of a specific statement, that such names are exempt from the relevant protective laws and regulations and therefore free for general use.

The publisher, the authors and the editors are safe to assume that the advice and information in this book are believed to be true and accurate at the date of publication. Neither the publisher nor the authors or the editors give a warranty, expressed or implied, with respect to the material contained herein or for any errors or omissions that may have been made. The publisher remains neutral with regard to jurisdictional claims in published maps and institutional affiliations.

This Springer imprint is published by the registered company Springer Nature Switzerland AG
The registered company address is: Gewerbestrasse 11, 6330 Cham, Switzerland

If disposing of this product, please recycle the paper.

To my family:
Sarah, Dane, Daisy, Michael, and ... Roshan

Other Books by Sadri Hassani

- Mathematical Physics: A Modern Introduction to its Foundations
- Mathematical Methods for Students of Physics and Related Fields
- Special Relativity: A Heuristic Approach
- From Atoms to Galaxies: A Conceptual Physics Approach to Scientific Awareness

A Physicist's Apology

In the 1920s, some great physicists publicized the falsehood that quantum physics was somehow connected to Eastern theosophy, and aside from Einstein and Planck, few notable physicists stepped in to correct the falsehood.

In the 1970s, a physicist wrote an inimically popular book that alleged a parallel between physics and Taoism, while other physicists did very little to challenge the allegation.

In the 1980s, a medical doctor fabricated "quantum healing," accusing quantum physics of having an affinity with the ancient religious medicine of India, and the physics community remained essentially silent.

In 2007, a panel on *The Oprah Winfrey Show* suggested that according to quantum physics, simply *thinking* about losing weight, making more money, and falling in love, you can become thin, wealthy, and happily married, and we physicists did effectually nothing to dispute that brazen nonsense.

As a concerned physicist, I, on behalf of my profession, apologize for all that and for things having gotten so out of hand that our most prestigious medical schools are now teaching our future doctors unsubstantiated alternative medical protocols mostly tied to Eastern theosophy, and billions of dollars of our precious resources are funneled into the teaching and practice of those protocols.

I have no excuse for my past acquiescence. In the future, however, I plan to devote a major portion of the royalties of this book to a fund earmarked for promulgating critical thinking and awakening the public from the stupor of aforementioned false beliefs. It is an ambitious task, and in light of the deluge of misinformation and apathy of the media, the task may seem impossible. But every song starts with a note, and this book is nothing but a single note. Let's turn it into a song! To see how you may help and to get more information, please go to quantaindistress.org.

Preface

One of the oldest and most neglected forms of misinformation and disinformation—much older than the alternative fact, the deep state, and the claim of the efficacy of ivermectin as a COVID-19 cure—is the untruth that there is some kind of a relation between quantum physics and the New Age favorites like meditation, yoga, shamanism, and a plethora of ancient Far Eastern methods of healing. Too many of the publications of this kind of falsehood have become best sellers and not enough has been done to expose the untruth. This dearth of exposure has helped turn the falsehood into a viral social disease with multifarious strains.

When I conceived of writing this book, we were living in a different world. In that world, it was inconceivable that the Supreme Court of a state would issue a ruling declaring that embryos created through in vitro fertilization should be considered children, and the governor of another state—in anticipation of banning it—attacks lab-grown meat as part of a "woke" liberal agenda. It was unimaginable for a weather reporter to get death threats for talking about climate change. In that world, a licensed medical doctor would never have claimed that gynecological problems were caused by having sex with demons and witches in dreams, and that scientists were cooking up a vaccine to prevent people from being religious, and still be allowed to go on national television to advertise hydroxychloroquine as a cure for COVID … and be backed by the sitting president.

The book was intended to be a—significant—corollary to the general theme of science literacy, aimed at alerting the public about the popular misconception that New Age practices like yoga, meditation, shamanism, and Eastern theosophies such as Taoism and Buddhism have something in common with quantum physics. My goal was merely to inject a small dose of critical thinking into the mind of the public as an antidote to a blind faith in the spurious scientific basis of New Age irrationality. Although I was aware of the indirect physical harm such irrationality could inflict upon society (the diversion of billions of dollars that could be used in auspicious scientific medical research to buying ineffective dietary supplements and alternative medical treatments comes to mind), I did not anticipate doctors opposing vaccination in the middle of a pandemic, threat on the lives of those doctors who recommend vaccination, and a president prescribing the injection of disinfectant into the bloodstream as a cure for COVID.

The star of the January 6, 2021, insurrection—the face that personified the anarchy and lunacy of the attack on The Capitol—was the shirtless howling man with a horned hat and a painted face. The man came to be known as the "QAnon Shaman" of the insurrection. The title was not figurative. He indeed was a QAnon conspiracist who believed in the New Age shamanism. He also stretched my hunch of the physical damage to society from a mere diversion of resources to the reality of an irrationality that has become the dangerous principles of a major political party in the USA. This kind of unthinkable, uninhibited, extreme, naked science illiteracy makes the severance of science from nonsense, of pop-spirituality from quantum physics—my corollary—exigent.

This book will show that the claim of a connection between quantum physics and yoga, meditation, shamanism, or any other ancient tradition or scripture exported to the West by the swamis of the East is baseless. Despite the disconnect, however, the allegation of a link with Eastern thought has a long history that goes back to the inception of quantum physics. The sheer length of this history has solidified the link and has made the mind of hundreds of millions of people—some highly educated—susceptible to belief in an alternate reality, not unlike the alternate political reality in which an alarming portion of present-day America lives.

In numerous interviews, ardent supporters of Donald Trump have expressed their sincere belief that the 2020 election was stolen from Trump and that Biden is not the legitimate president of the USA. A devotee of *The Oprah Winfrey Show* would travel to Brazil to have John of God, a third-grade dropout and a convicted serial rapist, cure her breast cancer, because Oprah featured him on her show. QAnon followers believe that John F. Kennedy Jr. will be resurrected and install Trump as president. Some anti-vaxxers are convinced that the coronavirus pandemic is a cover for a plan to implant mind-controlling microchips and that the Microsoft co-founder Bill Gates is behind it. A zealous follower of Deepak Chopra truly believes that mind rules over matter, that by intense meditation you can eliminate your pain, even become ageless, and that these claims are based on quantum physics, as Chopra continuously hammers the quantum-Ayurveda connection into the minds of his multi-million followers.

There may appear to be a world of difference between the gun toting right-wingers and the namaskar gesturing New Age leftists, but that difference pales in comparison with the very thing that they have in common: denial of reality and degradation of science. The alternate reality, in which Ayurvedic mind-body treatment of cancer trumps chemotherapy and other science-based protocols, also houses white evangelical Christian anti-vaxxers, whose dogmatic belief eclipses the safety of a public immersed in a deadly pandemic. The emergence of the word "*conspirituality*," the embodiment of which is the howling horned Shaman of January 6, heralds the alliance of conspiracists and the New Age spiritualists. And as the latter thrive by attaching themselves to quantum physics, we have reached a point in history where the unquestionable truthfulness of an exact science companions a seditious insurrection. That is very scary!

This book does not disprove, degrade, or attack any belief system. It respects the freedom of an individual to practice Christianity, Judaism, Islam, shamanism, yoga, meditation, voodooism, or any other article of faith. Its purpose is to lay bare the falsehood that quantum physics is somehow associated with Far Eastern theosophy and pop-spirituality. "But why single out Eastern thought? Why haven't you picked on Islam, Judaism, or Christianity in the book?", you may ask. The answer is simple: I have not seen or heard of a book titled *The Torah of Physics*, or *Quantum Hadith*, at the same time that there are hundreds of titles linking quantum physics with Eastern theosophy.

Falsehoods have to be exposed as they can lead to a sick mind; and the destructive power of sick minds believing in falsehood and alternate reality has been proven in history during more than fifteen centuries of pre-Renaissance Dark Ages, and now is jeeringly on full display in America. Showing the light to those who live in the darkness of an alternate reality is urgent but not easy, and this book does not pretend to accomplish that. It is only a small step in untangling from the exact science of quantum physics the second half of the framework of that alternate reality: "conspirituality."

Strategy: Several books on the quantum-physics-mysticism connection have had a scarring effect on the collective mind of the public. Excerpts from a sample of those books have been unsparingly scrutinized in this book, because to see how New Age gurus kidnapped quantum physics it is imperative to unapologetically dissect their work. A casual reader may not see the flaws in the pop-spiritualists' persuasive reasoning but may be made aware of them if the reasoning is vigilantly dissected and the subtle flaws are laid bare. Once you read this book and note the shallowness of modern gurus' ostensibly convincing arguments, you'll see the next New Age book in a different light because the reasoning and syllogisms in all New Age books have a striking resemblance.

Technicality: I thought a great deal about how to handle the technicality inherent in any physics discussion. In the end, I felt obligated to write Chaps. 9 and 10 to explain the *real* modern physics—as opposed to the fabricated modern physics that New Age gurus purvey—even though it requires the introduction of some technical concepts. These concepts do not use any math and have been mostly moved to the Appendix for ease of reading of those two chapters. I have included them in the Appendix so that readers, who want to be decidedly convinced of the genuineness of my arguments in the book and are willing to spend some time understanding the arguments, have access to them. The full comprehension of the technicalities is not crucial for appreciating those chapters. Two sections of these two chapters titled "The Third Stage?" in Chap. 9 and "The Taoist's Denial" in Chap. 10 are absolutely non-technical and highly recommended readings.

It gives me great pleasure to acknowledge Dane Hassani, MD, for reading the entire manuscript and giving invaluable comments and suggestions for improving the book. Veena Korah, MD, also read the manuscript and suggested alterations, for which I am very grateful. I also thank Daisy Hassani, MD, for some crucial remarks concerning the operation of the website of the book. Last but not least, I want to thank my wife Sarah for putting up with me for over half a century, giving wise advice for the overall concept of the book, and placating me while I was writing it. Needless to say, the ultimate responsibility for the content of the book falls on me. Finally, I would like to express my appreciation to my editor Dr. Sam Harrison for taking up the project and for his enthusiasm in bringing the book to fruition.

Urbana, IL, USA   Sadri Hassani

Contents

1	**Prologue**	1
2	**From Myth to Philosophy**	9
	Philosophy Is Born	12
	Philosophy is the *Opinion* of the Philosopher	13
	Spirit Is Born	14
3	**Sins of the Fathers**	17
	Eastern Theosophy in Western Philosophy	19
	Rise of Mysticism in the West	20
	Eastern Theosophy Mars Quantum Physics	22
	Sins of the Fathers	24
	Admission of Guilt	29
	Einstein on Mysticism	31
4	**How Weird Is It?**	35
	Certainty of the Probable	35
	Probability and Macroscopic Weirdness	38
	Conscious Coins?	39
	Resistance to Probability	42
	Quantum Tunneling	43
	Non-locality	45

5 From Duality to Mysticism — 49
A Gift from Nature — 50
From Uncertainty to Complementarity — 52
The Double-Slit Experiment — 55
Explaining the Double-Slit Experiment — 56
The Myth of Duality and Complementarity — 58
Objectivity Is Alive and Well — 61

6 Quantum Consciousness Crosses the Atlantic — 65
Oppenheimer's Mystique — 65
Dyson's Consciousness — 66
Wigner's Consciousness — 68
Weyl's Consciousness — 70
Wheeler's Consciousness — 70
Experimenter's Consciousness — 74
Admission of Guilt — 76

7 Eastern Plague of the Sixties — 79
Institute for Advanced Mysticism — 80
Dancing Shiva and Wu Li Masters — 81
Reality of the Unreal — 82
Conscious Photon? — 84
Classical Physics and Its Umbilical Cords — 86
Perversion of $E = mc^2$ — 88
Infesting Modern Physics — 91
Canine Mysticism and Field Theoretic Sanskrit — 94

8 The "Quantum" Healer — 97
Mysterious Disappearance of the Maharishi — 97
Quantum Oinking — 99
"Quantum" Healer's "Quantum" Theory — 100
You and the Universe — 103
Consciousness: The Cure-All of All Questions — 109
Consciousness: God in Disguise — 110
Quackia: Quanta of Fairies — 112
"Quantum" Healer Attacks Physics — 116
Where Does Consciousness Come From? — 119
A Modern St. Augustine — 123

9	**Basic Building Blocks**	125
	Science Cries for Reductionism	125
	Ordinary Matter	127
	Extraordinary Matter	129
	The Language of Nature	131
	New Physics Befriends New Math	133
	Quarks	136
	The Third Stage?	137
	Leptons	140
10	**The Standard Model**	141
	Symmetry and Gauge Theory	142
	Spontaneous Symmetry Breaking	144
	Electroweak Nuclear Force	145
	The "Charm" of Mathematics	146
	Higgs Boson	147
	Confinement and Strong Nuclear Force	149
	The Taoist's Denial	150
11	**Epilogue**	155
	Science Is Detached from Scientist	156
	Don't "Make Sense"!	158

Appendix — 163

- Homeopathy — 163
- Coin Probabilities — 165
- The Double-Slit Experiment — 167
- Sizes and Distances of Moon and Sun — 168
- French Teenager Invents New Math — 170
- On Spin — 171
- The Eightfold Way — 172
- The Quark Model — 174
- On Gauge Theory — 175

Notes — 179

Bibliography — 197

Index — 199

1

Prologue

In a series of episodes aired on *The Oprah Winfrey Show* in 2007, Oprah talks about the then-new sensational phenomenon known as *The Secret*, a movie by the Australian film producer Rhonda Byrne, who later wrote a book with the same title, which, due to Winfrey's enthusiastic endorsement, has sold over 30 million copies worldwide and has been translated into 50 languages. *The Secret* maintains that by *thinking* about losing weight, making more money, and falling in love, you can become thin, wealthy, and happily married. In one episode, Rhonda Byrne is joined by four "teachers"—other well-known self-help gurus who had chosen to disseminate the idea, much like the disciples of a prophet—in a speciously scientific discussion of the law of attraction, magnetic power, energy, frequency of mind vibration, and the vibration of the universe. All these buzzwords are the overture to the selling point of the conversation in which the author of *Chicken Soup for the Soul* proclaims, "If you go to quantum physics, we realize everything is energy."[1]

Marianne Williamson, former Democratic presidential candidate, designates "quantum realm of possibilities" as the source of "the good, the true, and the beautiful," and a solution to slavery, disenfranchisement of women, and segregation. She resorts to quantum physics to assert that "as our perception of an object changes, the object itself literally changes," and that to change the world all we have to do is change our mind about the world. With this premise, would-be President Williamson's solution to world problems—the conflict in the Middle East, Africa, and South America, the tension between Iran and the US and between North Korea and the West, racism and unsustainable income inequality in the US, a worldwide pandemic, …—is only a meditation away; and she has quantum physics to back her up![2] This is not a far-fetched claim that I have farcically concocted. When Hurricane Dorian was dashing toward Florida, Williamson advised her followers to stop it with their minds, insisting that she's not crazy for doing so.[3]

The rush of gurus to the West in the 1960s, combined with the atrocities committed in the Vietnam War, popularized the "peaceful" Eastern thought—symbolized by a smiling monk in a namaskar gesture—in the collective mind of the West. Several physicists, then graduate students or post-docs at universities in California, exploited some uniquely strange features of quantum physics to inject in it New Age ideas, which were growing rapidly on campuses all over the U.S. The appearance of bestsellers like *The Tao of Physics* by Fritjof Capra and *The Dancing Wu Li Masters* by Gary Zukav was the outcome of discussions ranging across such topics as philosophy, Hinduism, Buddhism, consciousness, parapsychology, and the paranormal in regular meetings in San Francisco and The Esalen Institute in Big Sur, California.

One of those gurus, Maharishi Mahesh Yogi (the Beatles guru), was in search of ways to make his signature spiritual practice, transcendental meditation, appear scientific. And when he found a keenly interested fellow countryman among his disciples by the name of Deepak Chopra, who had a medical degree, he was overjoyed. In one of their numerous meetings, the Maharishi instructed Chopra to "explain, clearly and scientifically," how certain meditation techniques work in healing diseases.*

With the misleading connection between quantum physics and Eastern thought having been popularized, Deepak Chopra's modus operandi became clear. All he had to do now was to stretch the Eastern thought to medicinal meditation. In his book, *Quantum Healing*, he concocts a sophomoric narrative that he calls "quantum physics" to connect Ayurveda, the mind-body medicine of ancient India, to science. The book added the concrete flesh of medicine to the earlier physics-Eastern-thought connections, which provided a mainly abstract philosophical skeleton. As a result, there occurred a revolution in "quantum" snake-oil medicine and self-help practices. Today, you can hardly find a pop-spiritualist who does not use the word "quantum"—or its siblings like energy, field, vibration, frequency—in conjunction with the basis of their practice.

The harmless mushrooming of yoga[4] centers—to the extent that the United Nations proclaimed 21 June as the International Day of Yoga—hides a malignant potion that seeps furtively into the mind of the public: the potion

---

*This is not unlike a rabbi instructing a scientist in his congregation to "explain, clearly and scientifically," how Moses parted the Red Sea.

of modern superstition. Unlike its old counterpart, which was weakened by the advancement of science, modern superstition, a prominent component of which is various forms of Eastern theosophy, purports to be *based* on the exact science of quantum physics. Google "meditation and quantum physics" or "yoga and quantum physics," and you'll get a dizzying number of sites attempting to demonstrate the association between the two. The association does not stop with meditation and yoga. Anything that is related to the ancient "wisdom" of East Asia becomes associated with hard sciences. Type acupuncture, chakra, Qi, Ayurveda, ... after "quantum" in a search engine and discover the hundreds of sites that try to convince you that they are all connected to quantum physics.

The benign but faulty association of New Age practices with hard science will likely subject their practitioners to *malign* falsehoods. If quantum physics supports ancient wisdom of the Far East and its noninvasive medical practices, then maybe the inhumane and cold-blooded surgical practices of the West, including the harsh vaccination attack on children, ought to be avoided and replaced with natural organic healing practices of the East. The mawkish Eastern-tainted "organic" becomes the antidote of fiendish Western-tainted "surgical," "chemical," "nuclear," and "genetically modified."

The crack that starts at one corner of reality spreads rapidly to its four corners. The step coming after—or in conjunction with—choosing Eastern theosophy over science is selecting falsehood over fact and fringe sources of information over mainstream media. The emergence of the phrase "alternative fact" in our time and the burgeoning of meditation and yoga centers, anti-vaccination movement, outbreak of measles, climate-change denial, and conspiracy theories cannot be dismissed as coincidences.

When ostensibly opposing groups attack science, they will, in all likelihood, wind up together. The same force that repels them from science attracts them toward each other. Eastern theosophy has delivered a subtle but heavy blow to science by mystifying quantum physics. In contrast, conspiracy theorists' attack on science is brazenly direct. In America, whose population, according to the noted cultural historian Richard Hofstadter, is markedly anti-intellectual,[5] the distrust of the elite easily translates into a distrust of the scientists "who think they know better than us, the ordinary folks." Steve Bannon's diktat that the infectious-diseases expert, Dr. Anthony Fauci, should be beheaded and have his head put on a pike outside the White House is not an aberration; it is the sentiment of millions of Americans.

In such a setting, the enemies of the enemy become friends: the hard-core advocates of alternative medicine, meditation, yoga, and quantum-based spirituality unwittingly find an ally in those who believe that all mass shootings are staged by Democrats and gun-control groups to take away gun owners' guns. Spirituality and conspiracy merge to form *conspirituality*.[6] This is not just a conjecture. The notorious QAnon conspiracy group, the group that believes in a cabal of Democrats and Hollywood celebrities participating in sex trafficking and baby blood drinking, has attracted dozens of celebrity yoga teachers and New Age influencers.[7]

The effect of social diseases like hate speech, neo-Nazism, and alt-right propaganda is, like a bodily disease, palpable and comprehensible: they abuse the right to free speech to sow hatred and incite violence in the mind of the uncritical public. In extreme cases of hard rhetoric of the speaker and low mental capacity of the listener, they even result in bodily harm and murder, as in Charlottesville, Va. where a brainwashed simpleton rammed his car into a crowd of peaceful counter demonstrators and killed a young woman.

New Age superstition, on the other hand, is a disease of the collective *mind* of the public, and much like a mental illness, it goes woefully unnoticed. In fact, because it is disguised as a harmless alternative to Western science, it has spread into a social mental pandemic. And just as some psychic abnormalities have somatic consequences, New Age superstition can have harmful *physical* repercussion. There is a fine line between choosing the wisdom of the Far East over science on the one hand and choosing the will of the God of the West to heal one's child over science-based medical treatment, on the other. More than one couple have lost their children for refusing treatment because of their Christian faith. And more than three hundred thousand individuals lost their lives too early for rejecting science-based medicine and embracing Ayurveda, herbal remedies, homeopathy, faith healing, and other pseudoscientific practices.[8]

An accurate barometer of the collective intellect of the public is the nature of the institutions of higher learning and what is taught and researched in them. The glory of the genius of Hellenistic Greece shone through the windows of academies, lyceums, and the Library of Alexandria where astronomy and geometry were pursued with joyous vigor and arduous rigor. The darkness of the Middle Ages emanated from monasteries and Inquisition where the Holy Book was the only source and authority on information. The feudal lords donated enormous resources to the intellectuals of that period—monks, priests, bishops, archbishops, and popes—to educate the public on their version of the teachings of the Bible. Any dissent would be crushed by the authorities, first in the Holy Inquisition, and if persistent, in the torture dungeons.

Those purveyors of New Age spirituality who hold enormous political power and wealth, pour millions of dollars into the pockets of intellectuals promoting pseudoscience to do "research" in centers and departments in credible scientific institutes and universities created under the pretense of evaluating the efficacy of ancient practices of the Far East. Any dissent would be crushed monetarily.

The National Center for Complementary and Integrative Health was established by Congress in 1992 under the strong backing of Sen. Tom Harkin (Democrat, Iowa) because he believed in the curative powers of bee pollen. The ostensible mandate for NCCIH—called Office of Alternative Medicine at the time and a little later changed to National Center for Complementary and Alternative Medicine—was to evaluate and determine the efficacy of alternative medicine. But Dr. Bernadine Healy, then the director of National Institutes of Health—in which OAM was to be an Office—and many of the outstanding medical experts she supervised, vigorously opposed the new office. They felt that such evaluations could be best and more objectively handled by the existing NIH structure. However, the wish of the Senator prevailed and ten years and a billion dollars later we are informed that

> OAM, the NCCAM and their advisory committees have been loaded with New Age gurus like Andrew Weil, assorted mystics, quacks – like the one that treated Harkin's allergies with bee pollen – as well as various hangers-on who apparently think that "placebo" refers to one of the Three Tenors. Indeed, a recent director of the Center, Wayne Jonas, proudly listed in his resume the authorship of a book called "Healing with Homeopathy".*[9]

It has been more than thirty years since OAM was established and over three billion dollars[10] have been appropriated to OAM, NCCAM, and NCCIH. Yet a simple question like "Does acupuncture—or any other alternative medical protocol—actually work?", a question, finding the answer to which was part of the mandate for the creation of OAM, has not been answered. A typical research paper funded by the Center ends with the grant-money-black-hole statement "further study is needed to ...."

Now compare the aforementioned alternative medicine mammoth with the websites Quackwatch[11] and Science-Based Medicine[12] which educate the public about the quackery of alternative medicine and other pseudoscientific practices, and rely on small donations to remain afloat and to pay for countering libel and defamation cases brought about by quack doctors and pseudoscientists of the kind who get grants from NCCIH to do "research"

---

*See page 163 for homeopathy.

in acupuncture, Qi, homeopathy, Ayurveda, and many other unscientific alternative medicine practices. They are indeed crushed monetarily by the wealthy and powerful.

The Dark Ages did not spontaneously appear out of a vacuum. It was the culmination of a gradual process that started with the replacement of Greek ideology of science and rationality with the Roman ideology of militarism and brutal gladiatorial entertainment. That today's universities and research institutions, including Harvard, Yale, Mayo Clinic, and National Institutes of Health, started to embrace acupuncture, Ayurveda, chakra, Qi, and other alternative medicine protocols at the turn of this century—an endeavor in which quantum physics, because of its false association with Eastern mysticism, played no small part—paints a dark picture of what our institutions of higher learning will look like by the end of the century.

How did quantum physics become a companion of Eastern mysticism? What is it about quantum physics that attracts modern gurus? Did the attraction start in the 1960s when Eastern culture and folklore flooded the West, or was there any precedence? And last but not least, does the *science* of quantum physics support Eastern mysticism as New Age mystics claim? These are important questions that go to the core of the dilemma of a society entrenched in science illiteracy, which arguably is the source of the willingness of the public to spend tens of billions of dollars on alternative medicine and dietary supplements;[13] of the spread of anti-vaccination movement and the consequent surge of COVID-19 among the unvaccinated; of the denial of the climate change exemplified by Sen. James Inhofe (Republican, Oklahoma) showing a snowball on the Senate floor to disprove the existence of global warming; of the attack on renewable energy epitomized by Donald Trump arbitrarily claiming that windmills cause cancer; of the governor of Texas blaming Green Energy for the collapse of unregulated, free-market, isolated Texan energy grid; and of the transition of conspiracy theories from the fringe to mainstream media.

Tracking down the roots of the false marriage between quantum physics and the mysticism of New Age gurus requires investigating the connection between philosophy and physics, the socio-political and philosophical developments in the nineteenth century, the impact of science on Western religions during the eighteenth and nineteenth centuries, and the intellectual environment of post WWI Europe. In the course of this investigation, it will be discovered that in an unfortunate coincidence, all founders of quantum physics had an affinity with Eastern theosophy and imparted their worldview on the foundation of quantum physics. Although attempts had been made earlier to associate other areas of modern physics—notably relativity theory—with mysticism, the attempts were not as successful.

Why quantum physics? Because quantum physics is unique in that it has certain strange features such as non-locality, or "spooky action at a distance" as Einstein called it, (that the measurement of a quantum particle in New York can instantaneously influence the outcome of the measurement of another particle in Paris);[14] or its introduction of probability at the most fundamental level. And these features defy explanation outside of mathematical reasoning. However, the founders, unprepared for accepting the spookiness of their own creation, could not swallow this new reality. Instead, they searched for something outside of physics and mathematics that "made sense" of the peculiarity of quantum physics, and Eastern theosophy—and a Western philosophy akin to it—seemed to be a good choice. Once the stigma of Eastern theosophy was attached to quantum physics by the founders themselves, it was next to impossible to detach it. Even the fierce opposition of giants like Einstein and Planck did very little to remove the stigma.

The architects of our civilization—scientists, mathematicians, writers, composers, poets, artists—share many of the same kinds of strength and weakness that we possess. Outside their areas of expertise, they are quite ordinary characters, who can be poor judges of politics, religion, morality and philosophical outlook. Newton believed in a 6000-year-old earth; Fourier, one of the greatest mathematical physicists of the nineteenth century, was a close friend of Napoleon's and compliantly witnessed his war atrocities; Einstein encouraged President Roosevelt to initiate the development of atomic weapons, an act which he later regretted; Linus Pauling, winner of two Nobel Prizes (chemistry and peace), was the originator of orthomolecular therapy, a dangerous alternative medical procedure; James Watson, the co-discoverer of the double helix nature of DNA is a racist; many great physicists participated in the Manhattan Project, which they later rued. But these mistakes are not made right because of the science of their makers, just as the science is not made wrong because of the mistakes of its discoverers. *It is the message that counts not the messenger.*

The physics community has, in large part, been publicly silent on the mysticism of the great physicists who founded and developed quantum theory and the damage their mysticism has brought on society. Perhaps it is because the scientific contributions of luminaries like Niels Bohr, Werner Heisenberg, and John Wheeler to our civilization are too valuable for their originators to be tainted with mystical beliefs. I myself had to struggle for decades to defend both physics and physicists while trying to convince my students in my science literacy courses that New Age gurus were spurious. I was blindly goaded by the popular writings, from which I taught my courses, into believing—and having my students believe—that there was an element of consciousness in quantum physics and that the observer had a role in creating the object being observed. As long as my students found the ideas "cool," I was encouraged to teach them.

Gradually, as I examined the philosophical writings of the founders of quantum physics, I came to the bitter realization that the root cause of quantum quackery lay in their beliefs and the dissemination of those beliefs to the public. However, until recently, my reverence for the founders was so strong that I could not bring myself to identify *them*—rather than the New Agers who simply and earnestly quote them—as the source of the abuse of quantum physics in the self-help industry. I kept reading critiques of the latest self-help bestsellers and cheered the critics' denunciation of modern gurus, rightfully portrayed as ignorant about the real quantum physics. I was, however, disappointed to see the critiques either defending the founders or at best remaining silent about them.

To remove the stigma of mysticism from quantum physics, we have to separate physics from the physicist—the message from the messenger. And no one can do this but scientists and science educators. They have to publicly draw a clear line between great physicists' contributions to physics and their mystical assertions—and, yes, criticize them for the latter.

As I was writing this book, I felt like the rope in a tug of war that was pulled on one side by my respect, admiration, and high regard for great physicists who created and contributed to quantum physics, and on the other side by my responsibility, as an educator, for informing the public that those very physicists were the unlikely source of the abuses to which quantum physics is being subjected by the exploitative New Age gurus. In the end, the rope was pulled to the side of the latter.

2

From Myth to Philosophy

Human knowledge is rooted in the murky origins of stories of creation. The actors in the drama of the formation of the cosmos possess a might and an intelligence that eclipse those qualities in even the strongest and brightest human beings. They could send a blinding flash to destroy a forest; roar in a deafening thunder; puff a storm that uproots trees, animals, and humans into the air like aimless bees and birds caught in a high wind; and dump water into a river to drown an entire civilization.

In the crowded habitat of gods, incredible events, filled with rivalry and cruelty lead to the creation of the world:

- A mother is beheaded by her four hundred children only to give birth to yet another child who decapitates all his siblings, including their leader, whose head flies into the sky to become the moon.
- A water beetle, desperate to find land in the infinite ocean, dives deep in the water and brings a speck of mud, which grows and grows until it becomes the Earth.
- An egg is born from of a universal being and the universal being is born again from the egg. The new "universal being" sacrifices himself to create animals, the sun, moon and stars, the air, the sky, the heavens, the earth.
- A water wagtail descends onto the watery world below, flutters over the waters, splashing water aside, and then packing patches of the earth firm by stomping them with his feet and beating them with his tail until habitable islands, big and small are raised to float upon the ocean.
- The daughter of the sky descends to the waters and becomes pregnant, gestating for a very long time but not being able to give birth, until a duck, seeking a resting place, flies to the knee of the pregnant girl and lays its eggs.

As the bird incubates her eggs, the girl is burned by the incubation heat and jolts her leg, dislodging the eggs, which then fall and shatter in the waters. Land is formed from the lower part of one of the eggshells, while sky forms from the top. The egg whites turn into the moon and stars, and the yolk becomes the sun.

Gods' "chosen" people on Earth are as fable-like as the gods themselves:

- A good shepherd who becomes king and rules in his first term for three hundred years, in the second term for six hundred years, and in the third for nine hundred years.
- An old man who is tasked by a god to abandon his worldly possessions and create a giant ship of solid timber and bring his family and relatives along with baby animals to save them from the oncoming flood, which is to wipe out everything not on the ship. … As the waters eventually recede and the old man sets all the animals free and makes a sacrifice, the god comes, and because the old man remained loyal and trusting of him, the god lets the old man and his wife live forever.
- A man who is given enough strength to slay a lion with his bare hands, massacre an entire army of the enemy using only the jawbone of a donkey, and bring down the columns of a magnificent temple, killing himself as well as the entire population of the enemy.

Myths are our ancestors' articulation of answers to their questions about the phenomena transpiring in their daily lives and an expression of hope—a much needed therapeutic instrument in a violent and unpredictable world. Stories of creation, salvation, might of gods and ogres, and courage of demigods, as outlandish as they may sound, have their roots in the physical world.

- A generation witnesses a cataclysmic volcano that wipes out half the population of a village. In telling the story of the eruption to the next generation, the elders' similes and metaphors take on a life of their own. And in the absence of writing, after a few generations (in a multi-generational game of Chinese Whisper,[1]) the story turns into a battle between the Gods of the Below World and the Above World over a beautiful princess.[2]
- During a calamitous flooding of a river, a brave man risks his life and his boat to rescue his fellow citizens. The tale of this man undergoes a slight change from one generation to the next until it turns into a flood that covered the entire world. The boat becomes a huge ark, and the man turns into a superhuman chosen by gods to rescue the world and let to live for nine-hundred years. Then writing is invented. Once the story is written

down, it acquires a documentary truthfulness which defies any challenge and conceals its relatively mundane origin. Does the story of Noah and his ark have a physical origin?

In the winter of 1929, the English archaeologist Leonard Woolley came across an assortment of tablets written in cuneiform near the city of Ur on the Tigris-Euphrates plain. One particular collection of tablets was copied hundreds, perhaps thousands of times and had been found in places as far away as Canaan. The tablets told of an immortal man named Ut-Napishtum who survived a universal deluge sent by the gods to punish mankind. According to the legend, Gilgamesh, the Sumerian demigod, sought out and found Ut-Napishtum, who told Gilgamesh how the world had become evil and how he was commanded by the Babylonian god Ea to build an ark and to load the animals upon it in preparation for "the end of all flesh." Woolley referred to Ut-Napishtum as the Babylonian Noah, but since the tablets were written 1500 years before the Genesis, Noah is to be recognized as the Biblical Ut-Napishtum. Was there any physical evidence for the flood?

As the excavation team dug deeper, they reached a point where all signs of civilization disappeared. All that was there was sterile river mud. They seemed to have hit the bottom of Ur. "Somewhere near 2800 B.C, all traces of human artifacts had broken off. But something about the ground seemed out of place to Woolley. On a hunch, he began digging into the thick clay and instructed the workers to join him. Twelve feet deeper their spades broke through into a horizon of collapsed walls and broken pottery. Difficult to believe, but there it was: a deep, uniform, water-deposited stratum of Euphrates clay, posited between two levels of civilization."[3]

As fascinating and tantalizing as these stories were, they could not indefinitely escape the inquisitive mind of the posterity who became increasingly reluctant to accept far-fetched stories at face value. As the mind of humankind began to rationalize, the eye of humanity turned to the seat of gods. The Sun, the Moon, and stars must surely be the chariots on which gods are riding, thought the Einsteins among our Egyptian and Babylonian ancestors. And if we scrutinize the motion of these chariots, we may secure a window into the thinking of the makers of flood, storm, thunder, and lightning, and perhaps a prospect of appeasing the gods.

The word of the birth of astronomy in the Egyptian and Babylonian temples reached the nascent civilization to their west that was eager to devour the stockpile of knowledge accumulated and recorded over the course of eons of observations and measurements by the Egyptian and Babylonian priesthood.

Philosophy Is Born

The Greeks famously had their own gods and demigods, constantly conspiring, machinating, and campaigning against each other. Poets, storytellers, and playwrights took delight in narrating the occurrences on Mount Olympus. However, the "love of wisdom" that drove the Greek scholars to Egypt and Babylon and introduced them to the emergent fields of astronomy and geometry started to sow seeds of doubt in the narratives of gods and goddesses. The sages of ancient Greece began to question the bizarre stories of life on Mount Olympus: Don't believe in the words of the poets and storytellers. Can't you see how impossible it is to turn a woman into a tree? It defies sensible thinking to accept that Poseidon could strike a rock with his trident and cause a spring of water to gush forth from the ground, or that Athena could strike her spear on the ground and make an olive tree jump out of the earth. Has any of you ever seen anything like those far-fetched occurrences?

The oldest instances of "critical thinking" started to shape the oldest schools of Greek philosophy. The analytic mind overpowered the illogical poetry about gods and titans. Along these lines, Socrates taught the youth—eager to learn, as was the intellectually delightful custom of the time—not to believe in the haphazard reality of gods and goddesses but put their minds to the rational experience of their world; to speculate about the heavens above and search into the earth below. Socrates paid with his life for the promotion of mind, in opposition to the sacred adoration of gods and devotion to the irrational beliefs that contradicted reason and logic.

The primacy of the mind, preached so powerfully by Socrates, took a unique place in Plato's philosophy. To Plato our senses are defective and cannot be trusted. He proposes that there is a real and perfect realm, populated by *Ideas* or *Forms* that are eternal and unchanging and constitute the character of the world presented to our senses.[4] And the only way to study reality is to resort to the organ in our body that can conceive the abstract *Ideas* or *Forms*: our brain.

The emphasis on the mind had two antipodal consequences. The first was Plato's focus on the role of mathematics in the education of the guardians, the elite philosophically trained class that governed his ideal state. During his dialogue with Glaucon in *The Republic*, Socrates—Plato's mouthpiece—enumerates harmonics, arithmetic, plane and solid geometry, and astronomy as the five subjects that a guardian has to master.[5] Solid geometry was just beginning to emerge and was at the forefront of scientific research at the time of the dialogue; and Plato wanted his guardians to engage in this abstract mathematical research.*

Greek cultural thirst for knowledge helped establish places in which the philosophical, mathematical, and scientific talent of the Greek youth could blossom. These places were not just the primary and secondary schools, but also institutions of higher learning such as Plato's *Academy*, Aristotle's *Lyceum*, and Epicurus' *Garden*, in which thinkers could exchange ideas, transfer knowledge, and create new knowledge. Greek aristocracy, themselves being a product of these institutions and an intelligent society in which they were raised, donated generously to the building of these centers. And Plato's emphasis on mathematics had a significant role in planting the seeds of rational and logical thinking in the minds of so many bright scientists, and nourishing politicians like Ptolemy I, who built the Alexandrian Museum, in which great scientists like Euclid, Aristarchus, Eratosthenes, and many others were trained.

The antipode to Plato's emphasis on mathematics was his disdain for experimental observation: if our senses are defective, any study of nature that uses those senses is also defective and should be avoided. While he glorified mathematics as an essential part of education and a window to *Forms* and *Ideas*, he chided his pupils who drew lines, circles, and triangles on sand to understand geometry better and encouraged them to create theorems and propositions purely from their minds. This may have hampered the advance of geometry at the beginning, but the later generations of geometers had no choice but to ignore Plato's advice and to rely heavily on such experimentation, thus making the rapid development of geometry possible and paving the way for Euclid to write *Elements*, his timeless treatise on geometry, almost a century later.

Philosophy is the *Opinion* of the Philosopher

When the mind—and only the mind—becomes the source of knowledge, haphazard things happen. The imagination of the forefathers of Greek philosophers created titans, gods, and goddesses and absurd stories chronicling their lives. Philosophy came to oppose such wild imaginations, but, as much as it

---

*Demanding that the guardians—counterparts of today's politicians—study solid geometry in Plato's time is like having our elected officials first master established advanced mathematical sciences, and subsequently do research in quantum gravity to qualify for public office. Would such rigorous prerequisites improve the critical thinking ability of today's conspiracy-smitten reality-denying politicians and inhibit their irrational views? Is the answer not obvious?

tried to tame the unrestrained activity of the mind by invoking reason and logic, philosophy itself became the driver of uninhibited speculations:

- Pythagoras believed in life after death and reincarnation in other forms of life such as a man becoming a dog after death.
- Heraclitus believed that everything constantly changes.
- Parmenides believed that nothing ever changes.
- Anaxagoras believed that everything is in everything and claimed that mind was the motive cause of the cosmos.
- Democritus held that there are indivisible bodies, or atoms, from which everything else is composed, and that these move about in an infinite void.
- Aristotle said there is no such thing as void.
- Plato said we should not trust our senses as gaining knowledge about the outside world.
- Epicurus said that our senses are the only way we can gain knowledge of the outside world.
- Kant proposed a *thing-in-itself* that is the cause of all things that we observe.
- Schopenhauer said that there is no such thing as *thing-in-itself*, and that our mind creates everything that we observe.

The diverse, even opposing, views sampled above prevail in philosophy throughout its history. Because the "reason" and "logic" of philosophers come from their unrestrained minds, in the final analysis, *philosophy is the opinion of the philosopher*. Opinions cannot be trusted as statements of truth, and no amount of debate, argument and counterargument, even among the most eminent philosophers, can bring out the truth.[6]

Spirit Is Born

As the skyward eye of humankind detected first gods and goddesses and then regular and predictable motion of objects such as the Sun, Moon, and planets associated with them, there seemed to be another *invisible* presence catching the eye of humanity as it looked around earthward. Our ancestors felt this presence every time they walked alone at night and sensed some *thing* following them; or when they could clearly hear someone speak to them in the howling of the wind; or when a ferocious beast appeared to attack and devour them right before they found themselves sweating after a nightmare; or when they bent over a tranquil pond to have a drink only to encounter another person approaching them from under the water.

2 From Myth to Philosophy

This unseen presence, invisible *spirit*, was so uniquely human and humanly universal that it transcended geography and culture. In the South Sea Islands, they called this mysterious force *mana*; the Latins experienced *numina* in sacred groves; Arabs felt that the landscape was populated by the *jinn*;[7] in the Far East all material objects, living or nonliving, possessed the life force or consciousness called "Vijnana" while Brahman was the "Supreme Spirit" who was male, female, even animal at the same time.

The universality of the telluric "spirit" went hand in hand with its regional non-uniformity. In Egypt and Babylon, the spirit became an unequal partner of the celestial forces that were believed to dominate the fate of the citizenry and its habitat and to whom sacrifices and prayers at magnificent temples were aimed. And when the tribes of Canaan inherited the Babylonian epic poetry of *Enuma Elish*, they perfected it into the narrative of an almighty God who lives in heaven, yet is present everywhere on Earth and is the ultimate cause of all things terrestrial and celestial.[8]

The monotheistic worldview was the fruit of a way of thinking that sought answers to earthly problems in the sky. The seeds of a single almighty god governing the fate of the world predates the Canaanite Yahweh. Babylonians believed in Ellil, the god of wind, air, earth, and storms and the chief of all the gods, the transcendent facet of Anu, the supreme God and prime mover in creation who was embodied by the sky. The Egyptian Ra, the Sun God, was one of the most important gods who ruled in all parts of the created world: the sky, the Earth, and the underworld; and in conjunction with the god Aten, Ra became the *only* almighty god of a monotheistic cult during the reign of Akhenaten.

Ironically, such reductionist focus on one celestial deity helped the progress of astronomy in Egypt and Babylon. However, reduction of objects of attention was only one of the two pillars of the birth of astronomy. The *regularity* of the motion of celestial bodies was also crucial in its development. The Sun and Moon always rose from the east and set in the west, and, aside from some seasonal variations, followed exactly the same path over and over again. This motion was so regular that Egyptians and Babylonians planned their daily activities around it. The solar calendar used in ancient Egypt points to the importance of the Sun. Babylonians were more fascinated by the Moon and chose its phases to guide their daily chores. But the Sun and Moon were not the only objects attracting the curiosity of the Egyptian and Babylonian priests/astronomers. The regularity of the journey of the five visible planets in the night sky was too fascinating to miss. The detailed record of this journey, once it fell into the hands of the Greek astronomers, lay the foundation upon which the post-Renaissance European scientists erected modern astronomy.

The fundamental marriage of *reductionism* and celestial *regularity* was at the heart of two antipodal developments: monotheism and science. Without the focus on the few celestial bodies that begged attention due to their regularity, the notion of an almighty who plans everything based on the infinite knowledge he possesses would have been next to impossible. By the same token, without the painstakingly recorded observation of the regular motion of the Sun, Moon, and planets by the priests of the ancient world, the Greeks would not have been able to advance astronomy, geometry, and trigonometry that became crucial in the development of science by the later generations of Greek and post-Renaissance mathematicians.

Contrary to Egypt and Babylon, where the terrestrial spirit was only a minor appendage to the celestial powerhouse, the Far East saw the telluric "life force" as the ubiquitous driver of all phenomena. There was no need to focus on a few objects in the sky and study them. The "force" was everywhere right here. And because no apparent regularity exists on the ground, the road to astronomy, geometry, and trigonometry was blocked. The holistic worldview, so cherished by New Age gurus, has always been a hindrance to science.

When the philosophical ingredient of focusing on the discernible regularity of a few things is lacking, everything becomes the center of attention: trees, rocks, animals, flowers, rivers, pebbles, humans … all have consciousness, all are both cause and effect, and therefore, all are equally worthy of examination. In this infinite confusion, science has very little chance of blooming. That astronomy—or astrology, which was the precursor of astronomy and survived until Newton discovered the mathematical formula for gravity and explained the planetary motion based on that formula—did not develop as fully in the Far East as it did in the West, is no coincidence. In the Brahmajala Sutta, Buddha speaks out against astrology and considers it lowly and unworthy of consideration. He chides ascetics and Brahmins who make their living by *such base arts* as predicting an eclipse of the moon or the sun and warns that those Brahmins will go astray.[9]

Ironically, thousands of years later, thanks to an unwarranted admixture of opinions and science, this haphazard way of thinking, which itself inhibited science, finds a companion in one of the most orderly and mathematically precise scientific discoveries of all time.

3

Sins of the Fathers

The word in Farsi for "foreigner" is *farangi* and for "foreign land," *farangestan*. They are the Persianized version of *French* and *France*. Their roots go back to the seventeenth and eighteenth centuries when Europeans intensified their interest in the Middle and Far East. France had a near monopoly of influence in Iran, so much so that Iranians identified French visitors as the sole representatives of foreigners. The fact that the founder of thermodynamics, Nicolas Léonard Sadi Carnot, takes his third given name from Saadi Shirazi, the great thirteenth century Persian poet and literary figure, indicates the mutual influence of East and West on each other. The East was fascinated by the technological advances of the West while the West searched for the wisdom of such ancient luminaries as Buddha and Zoroaster.

He was only 23 years old when Abraham Hyacinthe Anquetil du Perron set foot on the ship anchored at the port of L'Orient. Having developed a passion for Hebrew, Arabic, Farsi, and other languages of the East, he fancied a voyage to India, with the view of discovering the works of Zoroaster. Seeing no other means of accomplishing his plan, he enlisted as a common soldier in the Indian expedition which was about to start.

During his eight-year adventure and interaction with the native priests in India, Anquetil acquired sufficient knowledge of various languages to enable him to translate several texts, including the dictionary called the *Vedidad-Sade*. Returning to Europe in an English vessel, he spent some time in London and Oxford, and then set out for France. A year later, he began to arrange for the publication of the materials he had collected during his Eastern travels. In 1771 he published the *Zend-Avesta*, containing sacred writings of the Persians, a life of Zoroaster, and fragments of works ascribed to that sage. The work is considered the most authentic source of information on the religion and institutions of the great Persian prophet.

In 1801, Anquetil published his most influential work, a Latin translation from Farsi of *The Upanishads*, or "the greatest secrets," ancient Sanskrit texts that contain some of the central philosophical concepts of Hinduism. This translation became the authoritative source not only for scholars interested in Indian theosophy, but also for mystical cults and spiritualists that were beginning to mushroom across Europe.

The publication of *The Upanishads* contributed significantly to the infiltration of Eastern theosophy in Western thought, and consequently, to the essence of the philosophical appendage to quantum physics. It was not, however, sufficient for the mystical and consciousness-laden interpretation of quantum physics—an endeavor that was undertaken in other areas of modern physics but was not as successful as in quantum physics. Three factors had to converge to give rise to the association of mysticism with quantum physics: the esoteric infiltration of Eastern thought into Western philosophy, the exoteric upsurge of occultism and spiritualism in the West, and finally, the bewildering characteristics of quantum physics, not the least of which was its inability to predict anything more than the probability of occurrences of physical phenomena.

The strangeness of the probabilistic aspect of quantum physics can be, if not fully comprehended, then at least appreciated in the context of random occurrences in everyday life. If we observe that coins exhibit some strange behavior as a result of their randomness, and if we comprehend that such behavior can be explained by an exact mathematical formulation, then the strange behavior of quantum physics associated with probability at least, may not seem as strange (if we notice that such behavior can also be explained by exact mathematical formulations).[1]

But quantum physics has another kind of weirdness with no analogue in everyday life: it is the phenomenon of the "spooky action at a distance," whereby the measurement of a property of one member of a pair of particles influences the outcome of the measurement of the same property of the other particle, even if the two particles are separated by hundreds of miles. This phenomenon has been termed *entanglement* or *non-locality* and has enraptured quite a few mystics, who interpret non-locality as the manifestation of a universal consciousness with telekinetic power.

If the otherworldliness of quantum probability can be lessened by examining its analogue in the context of tossing coins, then perhaps the otherworldliness of "spooky action at a distance" can be lessened by noting that the same quantum laws that apply to quantum probability, also apply to quantum spookiness. John Bell, the late British physicist, did exactly that. He explained entanglement and non-locality within the context of the same mathematical formulation of quantum physics that explains quantum probability, and showed that they do not violate any physical law and need no extrasensory agents for their explanation.

Eastern Theosophy in Western Philosophy

As he was putting the finishing touches on his PhD thesis in 1813, Arthur Schopenhauer was introduced to Anquetil's translation of the *Upanishads*. He was so intrigued by the texts that he described them as the production of the highest human wisdom, and believed they contained superhuman concepts. The book always lay open at his table, and, as a rule, he read passages of the book before going to sleep. To Schopenhauer, the *Upanishads* were the most satisfying and elevating reading in the world. He called the translation of Sanskrit literature in Western languages "the greatest gift of our century" and predicted that the philosophy and knowledge of the *Upanishads* would become the cherished faith of the West.[2] And when Buddhism was introduced in Europe during the first half of the nineteenth century, Arthur Schopenhauer was delighted by the similarity it showed to his own philosophy. Having completed his major work, *The World as Will and as Representation* as early as 1818, he considered it an entirely new expression of the wisdom once taught by the Buddha. The book regards the world as having two manifestations: As Will, the world is as it is in itself, which is a unity; as Representation, the world consists of appearances, ideas, or objects, which is a diversity.

The Hindu dualism of Brahman and Atman with Brahman being "unlimited, unborn, not to be reasoned about, not to be conceived" and Atman being the true self,[3] has a striking resemblance to Schopenhauer's Will and Representation. This should come as no surprise, because by the time his book came out in 1818, Schopenhauer had been perusing the *Upanishads* for five years.

In Schopenhauer's philosophy, as in Buddhism and Hinduism, reality does not exist independent of thought. Mind and consciousness are the creators of the outside world and there is no distinction between the observer and the observed. The subject and object are unified in a single reality.

Rise of Mysticism in the West

One of the unintended consequences of the physics that began with Galileo and Newton in the seventeenth century was the eventual decline of the traditional western religions. Laplace's response, "Sire, I had no need of that hypothesis.", to Napoleon's remark that there was no mention of God in Laplace's *Mécanique Céleste*, and Nietzsche's climactic declaration—through Zarathustra's mouth—that "God is dead.", created a moral vacuum by the end of the nineteenth and the beginning of the twentieth centuries that could be filled only by a belief system that worshipped no supreme being. The filler turned out to be a salmagundi of spiritualism, esoteric western philosophies, and Eastern theosophy.

Spiritualism started as a movement based on the belief that the spirits of the dead existed and had both the ability and the inclination to communicate with the living. The afterlife was seen as a place in which spirits continued to evolve. Because of this evolution, spirits were believed to be more advanced than humans and could provide useful knowledge about moral and ethical issues, as well as about the nature of God. Mediums were individuals purported to be gifted with the ability to communicate—in sessions known as *séances*—with the spirits and learn about the knowledge they had gained in the afterlife.

Spiritualism and séances gained enormous popularity among European intellectuals in that period. A prominent supporter of the movement was Sir Arthur Conan Doyle, the creator of the legendary detective, Sherlock Holmes. Doyle also wrote non-fictional spiritualist works in which he tried to prove the existence of fairies.

A leading opponent of the spiritualist movement was Harry Houdini, the American magician, who, by replicating mediums' feats, tried to convince the public that séances were nothing but a showcase for magical tricks. Doyle and Houdini were friends, but Doyle refused to agree with Houdini. On one occasion, Houdini performed an impressive trick in the presence of Doyle, whom Houdini assured that the trick was pure illusion and that he was attempting to persuade Doyle not to endorse phenomena simply because he had no explanation for them. Doyle, nevertheless, refused to accept that, and instead, assumed that Houdini himself possessed supernatural powers.[4]

The spiritualist movement was not confined to the literary circles. The trickery used in séances was so impressive that even some well-known scientists started to believe in the paranormal. A medium who caught the curiosity of the scientific community was Eusapia Palladino. In 1905, she came to Paris, where Nobel-laureate physicists Pierre and Marie Curie and some of their fellow scientists attended her séances.

Five days before his accidental death in 1906, Pierre Curie wrote a letter to his friend Gouy in which he elevated what he saw in his last séance with Palladino to a level worthy of physical research: "There is here, in my opinion, a whole domain of entirely new facts and physical states in space of which we have no conception."[5] Despite this declaration, Palladino's trickery was disclosed when two professors from Paris University examined her in 1906 and concluded that she was a fraudster.[6]

The flood of mysticism ravaging Europe at the beginning of the last century eventually found its way into the mainstream science. In 1920, Arthur Eddington, a British astronomer, published a popular book on relativity entitled *Space Time and Gravitation*, in which he introduced the special and general theories of relativity to a non-technical audience. But Eddington went beyond a mere exposition of the science. He arbitrarily subjected some of the mathematical symbols in the theory of relativity to his own philosophical interpretation of matter, emptiness, light, and motion. Although Eddington admitted that his ideas were controversial, he did not hesitate to disseminate them—and more troublingly, publicize their association with relativity theory—to his uncritical readers:

> The theory of relativity has ... unified the great laws, which by the precision of their formulation and the exactness of their application have won the proud place in human knowledge which physical science holds to-day. And yet, in regard to the nature of things, this knowledge is only an empty shell – a form of symbols. It is knowledge of structural form, and not knowledge of content. All through the physical world runs that unknown content, which must surely be the stuff of our consciousness. Here is a hint of aspects deep within the world of physics, and yet unattainable by the methods of physics. And, moreover, we have found that where science has progressed the farthest, the mind has but regained from nature that which the mind has put into nature.[7]

This quote is an epitome of the syllogisms used by mystagogues to prove that their mysticism is based on science: first praise science, "the exactness of their application have won the proud place in human knowledge," then immediately point to its shortcoming in the world of mysticism, "And yet, in regard to the nature of things [mystical viewpoints], this knowledge is only an empty shell."

The idea of consciousness running through the physical world and mind creating the universe—"putting into nature"—is a recurring theme of all varieties of mysticism. While mystics distance themselves from religion and cuddle (modern) physics to prove their point, their mysticism is nothing but a religion in disguise. The statement above, that consciousness is "unattainable by the methods of physics," solidifies the religious nature of consciousness. Compare that to the often-quoted statement "God works in mysterious ways, but deciphering that mystery lies outside the scope of science."

Pick *any* New Age book, in which consciousness—universal spirit, organic energy, holistic energy field, ...—is claimed to be based on quantum physics. Pay attention to the explanation of the relation between consciousness and physics. You'll find that, after stripping the winded narrative to its bare bones, it reads something like this: "Physics has accomplished a lot, and consciousness is based on physics, but it cannot be explained by the experimental and theoretical methods currently available in mainstream physics."* The ruse works best when dealing with modern physics, because modern physics is outside the range of our sensuous experience. There is very little chance that the motion of a car, the heat of a flame, or the fall of an apple from a tree can be mystified. We have all experienced these (classical) phenomena without invoking mysticism. However, the process of staying young while traveling at high speed, the transformation of mass into energy, or the possibility of creating particles out of "pure" energy, is rife with mysticism. And Eddington, a mystic, saw the potential in these relativistic phenomena to give scientific legitimacy to his mysticism. However, his concoction was not sufficient to significantly quench the thirst of the public for the unification of science and the supernatural. For a "natural" and convincing unification, the public had to await the discovery of quantum physics.

Eastern Theosophy Mars Quantum Physics

As Europe was crossing the border between nineteenth and twentieth centuries, social, political, philosophical, and scientific upheavals were coming into being at a rate unprecedented in history. The sheer simultaneity of these upheavals portended an ostensible kinship among them. Physics, studying the fundamental concepts of space, time and matter, was more prone to this kinship than other branches of science. When he was a 23-year-old patent clerk in Bern, Switzerland, Einstein recruited philosophy student Maurice Solovine, and his friend, mathematician Conrad Habicht, to form a club mockingly

---

*See page 110 for a naked example.

called the *Olympia Academy*. The three met regularly in Einstein's apartment to read and discuss some physics, but mostly philosophy.

The social order of the time generated a keen interest in philosophy, politics, psychology, and mysticism among the generation of physicists coming immediately after Einstein. The 1920s, when quantum physics was born, were the years in which the memories of WWI were still fresh and the Bolshevik revolution, propelled by Marxian philosophy, had just succeeded in establishing the Soviet Union as an alternative socioeconomic system which was foreseen by the believers to eventually replace capitalism. They were also the years when Sigmund Freud advanced a revolution of a different kind, and spiritualists and their ostensibly miraculous séances caught the attention of the masses across Europe.

Imagine a gathering of intellectuals in Berlin, Paris, Zurich, or Göttingen in mid-1920s. What would be the topics of conversation there? WWI is over but its impact on society lingers. Lenin has died but his theories on Marxism, revolution, government, communism and socialism are being debated and stories from the Soviet Union that he helped create are the hottest news items. Freud's psychoanalysis and his emphasis on the animalistic sexual drives as the source of human development, especially his id—which, as an entity lacking any organization or discipline, has a remarkable resemblance to Schopenhauer's Will—ego, and superego trilogy are feverishly discussed. Plato, Aristotle, Kant, Nietzsche, Schopenhauer, and Marx are heatedly disputed, and voices raised in the discourse. And who can forget the incredible narratives about séances in which tables and chairs levitated and voices of the dead were heard. In such an atmosphere of intellectualism, mysticism, and political turmoil, it was natural for the founders of quantum physics, in their twenties, to be attracted to politics and philosophy.

But there were other reasons, more forceful than socio-political, that lured the founders to philosophy and mysticism. Quantum physics is a highly abstract theory with characteristics that outperform the weirdest phenomena introduced by relativity theory several years earlier. One example is *tunneling*, whereby a quantum particle located on one side of a *solid* (quantum) wall can be found on the other side. Other, considerably more mysterious than tunneling, are the notions of non-locality and entanglement. These are hopelessly inexplicable. Their very suggestion required years of attempting to digest quantum physics. Non-locality, for example, was the result of Einstein's desire to prove the incompleteness of quantum physics. He and two of his colleagues wrote a paper in 1935, almost a decade after Schrödinger's equation, in which they demonstrated the "spooky" non-local behavior of quantum physics. The fact that it took almost thirty years to prove—mathematically, and only mathematically—that quantum physics is non-local and there is nothing wrong with that, illustrates the degree to which such phenomena defy explanation.

The founders of quantum physics, however, could not swallow the weirdness of their own creation. They were looking for terrains of knowledge beyond mathematics and physics that could "make sense" of the strange behavior of quantum physics. This wasn't unlike Newton's "making sense" of the motion of the planets by having God place them at the right position, give them the initial push in the right direction, and let the mathematical law of gravity and the second law of motion take over subsequently. But "God [did] not play dice" as Einstein famously said about quantum probability. Besides, in the intellectual circles of the mid-1920s, God had already been killed by Nietzsche.

Buddhism and Hinduism, on the other hand, did not have an almighty God who was the cause and determiner of everything. As the indeterministic nature of quantum physics seemed to assent to the free will, which the deterministic Newtonian physics appeared to dismiss, the founders' philosophy of choice became Eastern theosophy and their philosopher of choice Arthur Schopenhauer, whose emphasis on mindless, aimless Will and its influence on reality through the process of objectification "made sense" of the probabilistic quantum physics.[8]

Sins of the Fathers

Whether through Schopenhauer or via a more direct path, the founders—Niels Bohr, Werner Heisenberg, Wolfgang Pauli, and Erwin Schrödinger—all developed a strong affinity for Eastern mysticism, and *regrettably tied their science to that mystical viewpoint.*

Niels Bohr talked about Buddhism, Taoism (founded by Lao Tse) and psychology and their role in quantum physics:

> For a parallel to the lesson of atomic theory regarding the limited applicability of such customary idealizations, we must in fact turn to quite other branches of science, such as psychology, or even to that kind of epistemological problems with which already thinkers like Buddha and Lao Tse have been confronted, when trying to harmonize our position as spectators and actors in the great drama of existence.[9]

With his *complementarity principle*—a philosophical statement derived from a mathematical inequality known as the *uncertainty principle*,[10] which limits our ability to measure certain quantities of a subatomic entity simultaneously—Bohr achieved an alleged reconciliation of physics and Eastern theosophy. The principle states that a subatomic particle (the object) has a dual nature—it could be localized and be a particle, or it could spread and act like a wave—and the two realities are mutually exclusive. It is the experimenter (the subject) who decides which reality to be attributed to the quantum entity. Therefore, the complementarity principle entangles the subject and the object, the epitome of Buddhism and Hinduism.

Werner Heisenberg was most influential in injecting Eastern theosophy in quantum physics. In his 1929 journey to the Far East, he had a long conversation with the Indian poet Rabindranath Tagore about science and Indian philosophy. He revealed that,

> After these conversations with Tagore, some of the ideas that had seemed so crazy suddenly made much more sense. That was a great help for me.[11]

And in Japan, he delivered a lecture in which he said:

> The great scientific contribution in theoretical physics that has come from Japan since the last war may be an indication for a certain relationship between philosophical ideas in the tradition of the Far East and the philosophical substance of quantum physics.[12]

Heisenberg is a giant of theoretical physics, and his genius has touched many branches of the discipline, and as a physicist, I have a lot of respect for him. So, it pains me to have to repudiate his statement above as an example of confirmation bias that plagues not only mysticism but sloppy scholarly research. The beliefs or "philosophical ideas" of scientists have no bearing on their science, as Heisenberg contends. We certainly do not conclude that Newton's discovery of gravity was "an indication for a certain relationship between" his belief that the earth was created 6000 years ago and his thought on gravity. Paul Dirac, who unified special relativity and quantum physics and consequently gave us anti-matter and founded the theoretical basis of the *Standard Model* and cosmology, was an atheist and detested mixing theosophy (of any kind) with physics.[13] Should we conclude that his atheism prompted relativistic quantum mechanics?

The intensity with which philosophy and religion dominated the thinking of the European physicists growing up in the early twentieth century is nowhere more transparent than in Heisenberg's recollection of the conversation he had with some young fellow physicists during the famous 1927

Solvay Conference. He recalls that during that conference, some of the younger members gathered in the lounge of their hotel to converse about religion and science. They were curious about the religious beliefs of Planck and Einstein, and the contrast between the two. While Planck firmly believed in a Christian personal God that was outside the realm of science, Einstein's God was the immutable laws of nature. Einstein's perspective—at least to Wolfgang Pauli, who was present in the gathering—allowed the unity of object and subject (although Einstein, an atheist himself, detested such unification and vehemently opposed any attribution of subjectivity to science). Planck, however, made sure that his faith (the subjective aspect of knowledge) was kept apart from his science (the objective side of knowledge). Heisenberg confessed that he himself did not feel happy about this separation and doubted whether human societies could live with so sharp a distinction between knowledge and faith.

Wolfgang Pauli shared Heisenberg's concern and pointed out that the separation of faith and knowledge would end in disaster. He argued that although at the dawn of religion all the knowledge of a particular community fitted into a spiritual framework, the advancement of society introduced knowledge that was in contrast to the old spiritual forms. Pauli saw this as a threat to the ethics and values of the society and found the solution in a spiritual framework where faith and knowledge, science and religion, object and subject are unified. He expressed hope in quantum physics:

> [the] very appearance of [complementarity] in the exact sciences has constituted a decisive change: the idea of material objects that are completely independent of the manner in which we observe them proved to be nothing but an abstract extrapolation. ... In Asiatic philosophy and Eastern religions we find the complementary idea of a pure subject of knowledge, one that confronts no object.[14]

Pauli's belief in Eastern theosophy was tied to his great admiration of Schopenhauer's philosophy. In one of a series of letters to Carl Jung about Jung's hypothesis of synchronicity, Pauli refers to an essay by Schopenhauer in which the philosopher postulates a power of sorts that connects everything in such a way that they meet at the appropriate moment.[15] Pauli likens Schopenhauer's thesis to Jung's "meaningful cross-connections" and concludes "This essay of Schopenhauer has exercised a lasting and fascinating effect on me, and he seemed to me to anticipate a future turn in the natural sciences."[16] Pauli's veneration of Schopenhauer is so great that he defends the pseudoscientific notion of extrasensory perception (ESP) because of the philosopher's belief in it: "Even so thorough critical a philosopher as Schopenhauer has regarded parapsychological [ESP] effects as not only possible, but as supporting his philosophy, which is indeed going far beyond what has been established by scientific empiricism."[17]

In one of the rare documents addressed to his sister Hertha, after his visit to Israel in the summer of 1957, Pauli writes, "I do believe that the natural sciences will out of themselves bring forth a counter pole in their adherents, which connects to the old mystic elements."[18]

In the introduction to the latest printing of *Quantum Healing*, Deepak Chopra plays a masterfully crafted trick that convinces more than a few of the most ardent skeptic of the connection between quantum physics and Far Eastern ancient wisdom. He quotes **Erwin Schrödinger** as saying "To divide or multiply consciousness is something meaningless. … In truth there is only one mind." Chopra goes on to claim, based on the quote above, that although it may seem that a person burning his hand on a radiator is separate from the person standing next to him, at a quantum level, this changes. "… you and I are imbedded in one mind, a cosmic intelligence that creates, governs, and controls reality. Where is this cosmic intelligence located?" Schrödinger comes to Chopra's aid again by "dissolving all the everyday barriers that keep us from seeing our cosmic status."

How does Schrödinger do this? Chopra lists some more quotes. First, the barrier of separate minds: "consciousness is a singular that has no plural." Next, the barrier of past, present, and future: "Mind is always now. There is really no before and after for the mind." Finally, the barrier between life and death: "Consciousness is pure, eternal and infinite: it does not arise nor cease to be. It is ever present in moving and unmoving creatures, in the sky, on the mountain, in fire and in air." Then Chopra pulls the Machiavellian trick out of his sleeve by announcing that the last quote does *not* come from Schrödinger, but from an ancient Indian text. If you, the reader, cannot differentiate between what the ancient Indian sages say about consciousness and what the founder of quantum physics says about it, is there any doubt that consciousness and quantum physics are one and the same?

Chopra hides the fact that all the quotes which he attributes to Schrödinger have their origins in the *Upanishads*. Schrödinger narrates his admiration of the Eastern theosophy in his biographical sketches, where he recalls that when the Emperor Karl abdicated and Austria became a republic, his life was affected by the breaking up of the Empire. Schrödinger had accepted a post as a lecturer in theoretical physics in Czernowitz and had already planned to spend all his free time acquiring a deeper knowledge of philosophy, having just discovered Schopenhauer, who introduced him to the *Unified Theory of Upanishads*.[19]

Schopenhauer's objectification by Will prompts Schrödinger to assert that the act of measuring the position of a particle actually *creates* the particle at that position. He complains that "the world of science has become so horribly objective as to leave no room for the mind and its immediate sensations."[20] Schrödinger recognizes the paradox of individuals having different minds while there is only one world, and finds the resolution of the paradox:

> There is obviously only one alternative, namely the unification of minds or consciousnesses. Their multiplicity is only apparent, in truth, there is only one mind. This is the doctrine of *the Upanishads*. And not only of the *Upanishads*. The mystically experienced union with God regularly entails this attitude unless it is opposed by strong existing prejudices; and this means that it is less easily accepted in the West than in the East.[21]

Schrödinger's infatuation with the mind and *the Upanishads* is so intense that he transcends the bounds of quantum mechanics and inserts the observer's mind into any kind of measurement. He asserts that

> the observer is never entirely replaced by instruments; for if he were, he could obviously obtain no knowledge whatsoever .... Many helpful devices can facilitate this work .... But they must be read! The observer's senses have to step in eventually. The most careful record, when not inspected, tells us nothing.[22]

By this assessment, neither the Moon nor the hotness of boiling water exists unless a conscious observer looks at the Moon or reads the thermometer measuring the temperature of the boiling water. It is a renunciation of matter almost as complete as Bishop Berkeley's "immaterialism" or "subjective idealism," which denies the existence of material substance altogether and instead contends that familiar objects like tables and chairs are ideas perceived by the mind and, as a result, cannot exist without being perceived.

To critics like Einstein and Planck, Schrödinger says:

> But some of you, I am sure, will call this mysticism. So with all due acknowledgement to the fact that physical theory is at all times relative, in that it depends on certain basic assumptions, we may, or so I believe, assert that physical theory in its present stage strongly suggests the indestructibility of Mind by Time.[23]

The assumption of nonexistence of objects without perception and the primacy of consciousness over matter is beautifully addressed by John Bell, who was not bound by any philosophical shackle:

The only 'observer' which is essential is the inanimate apparatus which amplifies microscopic events to macroscopic consequences. [I]t is a matter of complete indifference whether the experimenters stay around to watch or delegate such 'observing' to computers."[24]

Admission of Guilt

The groundless character of mystical beliefs of some great physicists shows up in the contradictory and indecisive statements they make regarding those beliefs. While they don't even have to defend their science because of its firm foundation in facts and verifiable observation, they waver when pressed on their mysticism to the point that they sometimes take back their beliefs.

Bohr Takes It Back ... and Forth: The flimsy character of the physicists' mysticism is most eminent in one of the strongest voices supporting the mystical interpretation of quantum physics, Niels Bohr. The famous 1927 Solvay Congress was convened to discuss the implications of the newly discovered quantum physics. It was also an opportunity for the founders of quantum physics to debate and disseminate mystical ideas that embodied Schopenhauer's philosophy and Eastern thought.

After this Congress, Einstein accused Bohr of tainting quantum physics with subjectivity and mysticism, both of which are incompatible with science. Bohr spent most of the rest of his life denying this charge and blaming it on misunderstandings. He admits that he incidentally pointed out that "even the psycho-physical parallelism as envisaged by Leibniz and Spinoza has obtained a wider scope through the development of atomic physics, which forces us to an attitude towards the problem of explanation recalling ancient wisdom, that when searching for harmony in life one must never forget that in the drama of existence we are ourselves both actors and spectators." He then continues to say,

> Utterances of this kind would naturally in many minds evoke the impression of an underlying mysticism foreign to the spirit of science; at the [Copenhagen Congress for the Unity of Science] in 1936 I therefore tried to clear up such misunderstandings. ... Yet, I am afraid that I had in this respect little success in convincing my listeners, for whom the dissent among the physicists themselves was naturally a cause of skepticism.[25]

Bohr kept flipping from mysticism to rationality and back. According to Heisenberg's recollections, as early as 1927 Bohr seemed to reject the hypothesis that quantum physics required a conscious observer. A few years later, Heisenberg asked him again about extending quantum physics to accommodate human consciousness. By then Bohr was less emphatic "… consciousness must be part of nature, or, more generally, of reality, which means that, quite apart from the laws of physics and chemistry, as laid down in quantum physics, we must also consider laws of quite a different kind. But even here I do not really know whether we need greater freedom than we already enjoy thanks to the concept of complementarity." Later, in his collected writings, he repeatedly distances himself from the consciousness hypothesis, labeling it 'mysticism':

> Still, the recognition of an analogy in the purely logical character of the problems which present themselves in so widely separated fields of human interest does in no way imply acceptance in atomic physics of any mysticism foreign to the true spirit of science.[26]

Bohr's vacillation between theosophical mysticism and scientific rationality, plus the opposition from others including Einstein and Planck, must be convincing evidence that his claim of a conscious entity enveloping the reality of quantum physics is false and has no affiliation with his science: Once observation proved the validity of Bohr's scientific work, he never tried to distance himself from it. There is nothing wrong with the vacillation per se. Einstein vacillated between introducing a cosmological constant in his field equation or not. In coming up with his eponymous equation, Schrödinger vacillated between incorporating special relativity or not. The difference between Bohr's vacillation and that of Einstein and Schrödinger is that the latter were *alternatives* to some mathematical statements. The former was the *negation* of a philosophical viewpoint versus its affirmation.

Schrödinger's Disclaimer: In the preface of his collection of essays on life, mind, and matter, Schrödinger laments the specialization forced on science by the enormity of knowledge and yearns the days—perhaps in pre-Renaissance Europe—when the institutions of higher learning (universities) were named after the all-embracing (universal) knowledge. He points to the spread of multifarious branches of knowledge in the nineteenth century and contrasts the breadth of this knowledge with the limitation of a single mind and concludes that this creates a dilemma. Although the dilemma goes away once one abandons Schrödinger's faulty notion that there has to be some kind of universal knowledge that makes peace with the specialized knowledge accrued as a result of scientific advancement, Schrödinger, being also a philosopher, and a follower of Schopenhauer at that, cannot let go of the dilemma of his own creation. So, he finds himself obligated to address it:

> I can see no other escape from this dilemma ... than that some of us should venture to embark on a synthesis of facts and theories, albeit with second-hand and incomplete knowledge of some of them – and at the risk of making fools of ourselves.[27]

When Schrödinger concedes that he may be making a fool of himself, he is separating his physics from his philosophy and alerts his readers not to take his philosophical utterances seriously.

Einstein on Mysticism

The baselessness of mixing science with beliefs can be demonstrated by the persistent, multi-generational opposition of credible scientists to those beliefs. Dirac refuted the mysticism advocated by Heisenberg and Pauli in the 1927 Solvay Conference. Much later, Richard Feynman and John Bell opposed any mixture of philosophy or Eastern theosophy with quantum physics. But Einstein and Planck, two giants of the old generation, having lived through the peak of the "Theosophy and Spiritualism" movements and having felt the threat they posed to science, showed the strongest opposition.

Einstein, an atheist, was particularly annoyed by the infiltration of mysticism—which had become a cultural zeitgeist by the beginning of the twentieth century—into physics. Of this cultural craze, Einstein said,

> The mystical trend of our time, which shows itself particularly in the rampant growth of the so-called Theosophy and Spiritualism, is for me no more than a symptom of weakness and confusion. ... the concept of a soul without a body seems to me to be empty and devoid of meaning.[28]

Einstein had several famous dialogues with Bohr concerning the interpretation of quantum physics and the idea of wave-particle duality, in which Bohr, an advocate of the connection between quantum physics and Eastern mysticism would use the probabilistic nature of the wave and Heisenberg's uncertainty principle to convince Einstein of the complementarity principle and, by extension, quantum-physics-mysticism connection. The use of probability (of a *single* particle), however, muddles the very notion of duality. Einstein saw the flaw in Bohr's reasoning:

> The Heisenberg-Bohr tranquilizing philosophy – or religion – is so delicately contrived that, for the time being, it provides a gentle pillow for the true believer from which he cannot very easily be aroused. So let him lie there. But this religion has so damned little effect on me that despite everything I say
> not: E and ν
> but rather: E or ν;
> and indeed: not ν, but rather E (it is ultimately real).[29]

Einstein is referring to the alleged unity of energy E (a particle attribute) and frequency ν (a wave attribute) and declaring that there is no such unity and concluding that E is the real thing and a particle never behaves like a wave. The use of the damning word "religion" depicts Einstein's frustration with the rampant mysticism infesting the minds of Bohr, Heisenberg, and others.

Einstein's fight with mysticism was not just cultural, but also scientific and personal. As early as 1921, years before the birth of quantum physics, Arthur Eddington had already mystified Einstein's general relativistic (GR) field equation. British and international media picked up on the Eddington's popular but misguided interpretation of Einstein's GR and were intrigued by the companionship of the trendy mystical movements of the time and GR. When a journalist asked Einstein about this mystical relationship, his wife Elsa broke into laughter and said, "Mystical! Mystical! My husband mystical!" She was echoing Einstein's own reply to a Dutch woman, who expressed her fondness for his mysticism: "Mysticism is in fact the only reproach that people cannot level at my theory."[30]

Thanks to Eddington, the public associated Einstein with mysticism. The British Press had already broadcast the falsehood that Einstein subscribed to the idea that the outer world is a derivative of consciousness. Einstein reacted strongly to the idea in his press conferences and publications:

> No physicist believes that. Otherwise he wouldn't be a physicist. Neither do [Eddington and Jeans]. … These men are genuine scientists and *their literary formulations must not be taken as expressive of their scientific convictions.*[31] [Emphasis added]

It is fitting to end this chapter with repeating the sentence emphasized in the quotation above for every physicist who believes in mysticism because it lays bare the machination of New Agers who want to mix the message and the messenger:

- Bohr's literary formulation of complementarity as giving us freedom to assert that "consciousness must be part of nature, or, more generally, of reality" must not be taken as expressive of his scientific conviction.
- Heisenberg's literary formulation that "human societies could not live with so sharp a distinction between knowledge and faith" must not be taken as expressive of his scientific conviction.
- Schrödinger's literary formulation that "to divide or multiply consciousness is something meaningless and in truth there is only one mind; that consciousness is a singular that has no plural; and that there is no before and after for the mind" must not be taken as expressive of his scientific conviction.
- Pauli's literary formulation that "the natural sciences will out of themselves bring forth a counter pole in their adherents, which connects to the old mystic elements" must not be taken as expressive of his scientific conviction.

4

How Weird Is It?

"Laser precision" is commonly used to emphasize the exactness and extreme accuracy of something. It is also a contradiction in terms because laser is based on quantum processes which are fundamentally probabilistic. How can an object that is chaotic at its foundation become a hallmark of precision?

Certainty of the Probable

A(n ideal) coin is an epitome of randomness. No one can predict the outcome of flipping a fair coin—in the same way that they can predict the rising of the sun from east, the fall of an apple from a tree, or the illumination of a light bulb upon the flip of a switch. More precisely, any prediction of the outcome of flipping a coin can be accurate to within the probability of the two possible outcomes: if you say "head" every time a coin is flipped, on average, you will be correct only half of the times. And if you say "tail," your prediction will be just as good … or bad.

Coin tossing is the simplest random event, and therefore the easiest to study, yet the conclusions drawn from such a study are very general. It is therefore worthwhile to get familiar with the basic properties of the outcome of tossing a number of coins, especially when that number is very large. The reason for interest in large numbers is that, in quantum physics, we ultimately deal with an astronomically large number of atoms and molecules.

There are two outcomes when you toss a single coin: head (H) or tail (T). So, we say that the probability of getting H is one-half or 50%. Similarly for T, but I will concentrate on H from now on because once we know the number of

H's, the number of T's can be determined. The number of possible outcomes in flipping two coins is four: HH, HT, TH, TT, and the probability of getting two heads is one-fourth or 25%; the probability of getting one head is two-fourths or 50%; and the probability of getting zero heads is one-fourth or 25%. Now toss three coins and, by enumeration, convince yourself that there are eight possible outcomes and that the probabilities of getting three, two, one, or zero head(s) are, respectively, 12.5%, 37.5%, 37.5%, and 12.5%.

I can go on to four, five, six, ... coins and enumerate the outcomes. However, the process becomes extremely tedious very quickly, and since we are interested in a large number of coins, the enumeration becomes next to impossible.* Fortunately, there is a formula that tells us how many times a certain number of heads show up when a given number of coins are tossed and what the probability of getting that many heads is. I am not going to do any mathematical calculation with that formula here. I intend only to show some graphs of that formula which shed light on the general features of the probability distribution.†

Figure 4.1 shows the probability distributions for 10, 100, 500, and 1000 coins. The horizontal axis shows the number of heads and the vertical axis the probability. Two features of the distributions are noteworthy. First, they all have a maximum at the number of heads corresponding to half of the coins: 5 heads for 10 coins, 50 heads for 100 coins, 250 heads for 500 coins, and 500 heads for 1000 coins. Second, the curve gets sharper as the number of coins increases. This second feature is responsible for some of the strange behaviors of probabilistic phenomena mentioned earlier. So, it is important to understand this feature well.

Look at the graph of 100 coins (top row on the right) and note that the probability of getting a number of heads less than 35 or greater than 65 is practically zero. This means that if you toss 100 coins many many times[1] and each time count the number of heads, you'll see a number between 35 and 65 almost all the time, i.e., only 30 of the possible outcomes (the number of heads between 35 and 65) show up, while the other 70 (the number of heads smaller than 35 or larger than 65) rarely show up. In other words, the

---

*For a "mere" 1000 coins, the number of possible outcomes is more than 1,000, ··· ,000, with the dots representing 295 additional zeros!
†If you are interested in seeing the formula and how it works, you may go to page 166 for details.

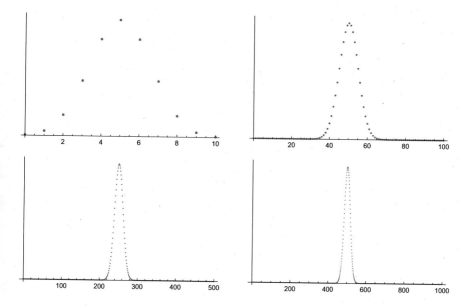

Fig. 4.1 Probability distributions for 10, 100, 500, and 1000 coins

nonzero* probabilities are concentrated in a 30% interval—of the horizontal axis—about the maximum, or 15% on either side of the maximum.

For 1000 coins (the graph at the bottom row on the right), the nonzero probabilities are concentrated between 450 and 550 heads or in a 10% interval (100 out of 1000) about the maximum, or 5% on either side. So, if you toss 1000 coins many many times and count the number of heads each time, only 10% of the possible outcomes (the number of heads between 450 and 550) frequently show up, while the other 90% (the number of heads smaller than 450 or larger than 550) rarely show up.

For larger and larger number of coins, the nonzero probabilities get squeezed in smaller and smaller percentage intervals about the maximum: for 100,000 coins the interval shrinks to 0.6% on either side of the maximum; for 100 million coins the interval reduces to 0.025% on either side of the maximum: any number of heads below 49.974 million or above 50.026 million practically never show up. Note how close 49.974 million and 50.026 million are to 50 million, the number of heads corresponding to the maximum probability.

---

*Actually, none of the probabilities, not even for the extremely unlikely one head or 100 heads, is zero. However, the majority of them are so small that we can assume they are zero. So, when I call the probability of getting a number of heads between 35 and 65 "nonzero," I'm cutting the "very smallness" at 35 and 65, and instead of saying the "larger-than-very-small" probabilities, I say the "nonzero" probabilities.

The upshot of this discussion is that as the number of tossed coins gets larger and larger, the chance of getting any number of heads ever so slightly different from half of the coins gets smaller and smaller. In other words, all probability is concentrated in the close proximity of the maximum. The probability that in tossing 100 million coins you get a number of heads between 49.974 million and 50.026 million is 0.99999985 or 99.999985%.

Probability and Macroscopic Weirdness

Coins, heads and tails, especially when there are too many of them, become too abstract. To get closer to reality, let's make the coins microscopic and paint their heads black and their tails white. Consider an experiment in which we put a very large number of these coins in a salt shaker, shake it and throw the coins on a table top. If the coins are spread evenly, you will see a shade of gray, much like the grainy black-and-white photographs in old newspapers. Because of the tendency of the coins to gather close to the maximum probability, we expect that the shade of gray that shows up will contain very close to half black and half white coins. Let's call the shade of gray with *exactly* half white and half black coins *perfect gray*. What are the odds of getting *anything but* perfect gray in our experiment?

To be able to quantify our experiment, suppose we have a trillion microscopic painted coins in our salt shaker. Since we expect the deviation from perfect gray to be small, we use an instrument that is very sensitive to shades of gray which are only slightly different from perfect gray. How sensitive? Mix half a gallon of white paint with half a gallon of black to obtain one gallon of perfect gray paint. Assume that our instrument is sensitive enough that it can distinguish between perfect gray and the shade obtained when one drop of black paint is added to a gallon of perfect gray. How does this imbalance between black and white paint translate into the imbalance of black and white coins?[2] The answer turns out to be 500,005 million black heads and 499,995 million white tails. Therefore, for the instrument to detect anything *but perfect gray*, the number of heads must be larger than 500,005 million or smaller than 499,995 million.

What are the chances that in a random toss of a trillion coins, the number of heads is larger than 500,005 million or smaller than 499,995 million? The answer is one minus the probability that the number of heads is between these two numbers, which turns out to be 0.99 · · · 9985 (replace the dots with 18 more nines),[3] and the number we are looking for is 0.00 · · · 0015 (replace the dots with 18 more zeros). That is, the odds are one in about sixty-seven trillion billion (67 with twenty-one zeros to its right) that we see a number of heads that is larger than 500,005 million or smaller than 499,995 million.[4]

Now let's go back to our salt shaker and ask for the probability that our very sensitive instrument detects anything but perfect gray. To appreciate the rarity of the occurrence of non-perfect-gray outcomes, suppose we have a machine that can fill the shaker, shake it really well, pour the content, and spread it on the tabletop 100,000 times per second.[5] In order for our sensitive instrument to detect any shade of black or white, we have to do the experiment at least sixty-seven trillion billion times. At 100,000 times per second, this requires 670 quadrillion seconds. Because there are about thirty million seconds in a year, we would have to wait more than 20 billion years. The age of the universe is 13.8 billion years! We can safely say that we shall *never* see anything but perfect gray when a trillion black and white coins are tossed.

This result is extraordinarily counter-intuitive: I cannot predict the outcomes of tossing one, two, or three coins. However, when tossing a trillion coins, I can predict with "laser precision" that the outcome is 500 billion heads and 500 billion tails (this is what we called "perfect gray"),* and to see any deviation from this prediction, I'll have to wait longer than the age of the universe. Trying to "explain" this paradox leads to absurdity, unfounded beliefs, and superstition.

Conscious Coins?

There are practically an infinite variety of shades of gray, from pure black to pure white. However, every time we perform the salt shaker experiment, the result is perfect gray. Why is it that out of an almost infinite variety of shades only the perfect gray shows up? Is there some kind of an "invisible hand" or a "life force" that purposefully arranges the coins so that the number of heads matches undetectably with the number of tails? Or perhaps the coins are "conscious" and when there are many of them, they consult with each other and "decide" to turn themselves up in a certain way! If we had not gone through the theory of probability, we would have suspected the intervention of some kind of outside consciousness.

Here is another intriguing example of probability and the "proof" that coins are conscious. On page 166, I have calculated the probability of getting 6 heads in tossing 10 coins. It is 20.5%. So, for 10 coins there is 20.5% chance that

---

*There is no *detectable* difference between 500 billion and either 500.005 or 499.995 billion.

60% of them (6 out of 10) turn up heads. Now toss 10,000 coins and ask: "What is the probability that 60% of them turn up heads?" The same formula that gave 20.5% for 10 coins gives $0.000\cdots00029\%$ (replace the dots with 81 more zeros). Now note that the 10,000 coins could be considered as 1000 groups of 10 coins. Each group has a 20.5% chance of showing 60% heads. But when they come together, this chance practically drops to zero. It is very tempting to say that the coins *know* that they are together, and the groups of ten tell each other "let's not show our heads!"

New Age gurus attribute consciousness to subatomic particles because, due to the probabilistic nature of quantum physics, subatomic particles, like coins, behave differently in large groups than individually. The difference is that we see the randomness of individual coins, but we are blind to the chaos of a single subatomic particle. And this difference renders any attribution of consciousness (or spirit, or soul, or Qi, or ...) to coins laughable. The realm of quantum physics, on the other hand, is beyond the reach of our experience, and the gurus can invoke consciousness when discussing photons, electrons, and other subatomic particles to "explain" their probabilistic behavior.

There is yet another example related to probability, whereby miracles and supernatural hands are assumed to be involved: a dream that actually comes true. Probability theory can explain such situations as well, albeit not as precisely as the behavior of coins. The images we see in our dreams are (sometimes distorted) copies of the most prominent scenes we see in our everyday lives. These images are, therefore, fairly limited in number, and they are shared by millions of people around the world. So, a nightmare involving a car accident severely injuring a relative should not be uncommon. Even if the probability of the coincidence of the dream and the real event is very small, with the population of the world surpassing eight billion, and the number of distinct dreams a typical person experiences in a lifetime being tens of thousands, it should come as no surprise to hear *genuine* stories of such coincidences.

One of the lessons we should learn from the theory of probability is that, no matter how small the chance of the occurrence of some very unlikely event, as long as the population of the sample is large enough,* there will be at least a few cases of that unlikely event. There is no need for introducing "miracles" and supernatural powers; the theory of probability and statistics can explain the unlikely event convincingly.

---

*See page 166.

We live by the rules of chance. Every time we step outside, there is a nonzero probability that we will get hit by a car, and another nonzero probability that once hit, we will die. Every time we go to the countryside, there is a chance that we will be struck by lightning. This chance increases in a thunderstorm, but even a sunny day can get cloudy quickly and provide a good condition for lightnings. As I am sitting in my living room reading a book, there is a nonzero probability that the small plane flying overhead will plummet into my house. There is about 0.03% chance that I will contract cancer this year, and once diagnosed, a 40% chance that I will die from it.

Unlike the tossing of coins, these probabilities cannot be calculated a priori, but a collection of data over many years can give a reliable magnitude for them. In fact, the figures of 0.03% and 40% were obtained from the data collected by the American Cancer Society over decades. These data consist of the number of new cases of cancer and the number of deaths caused by cancer every year. The data can give us a very good estimate of the probability. For example, over the last few years, the number of new cases of cancer has been stable at about 1.2 million per year in the US. Similarly, the number of cancer deaths has been equally steady: about half a million per year. Of course, diet, exercise, and annual check-ups can *reduce* the risk (from 0.03% to 0.02% or even 0.01%), but can neither eliminate it nor contradict the fact that contracting cancer is probabilistic.

Not only is the occurrence of the disasters above totally probabilistic, but also surviving them. The probability that, with all proper treatments, a cancer patient will survive five years after being diagnosed is about 60%. This survival probability decreases as the amount of treatment is reduced and/or the period after diagnosis is increased, but it does not become exactly zero. Even at the extreme case of no treatment at all, the probability of surviving for a long time and the complete disappearance of cancer is not zero. So, because the sample is the population of the Earth, quite a few cancer patients survive the deadly disease without any treatment.[6] The body itself has a mechanism of fighting the disease, and although most people's bodies lose the battle without outside help, the composition of the bodies of some patients is such that no outside help is necessary in the fight.

The survivor, of course, sees this completely differently. She looks at herself as one of the very few survivors of the disaster that kills hundreds of thousands of people. She thinks that she is one of the *chosen* few, whose survival must have a cosmic purpose. And if she has been seeing an alternative medical doctor or a faith healer (the chance of this is extremely high in cases of incurable diseases when conventional treatments reach the end of their road), she can be a priceless testimonial to the miraculous effectiveness of the alternative treatment or prayer.

Resistance to Probability

Quantum physics did not introduce probability to physics. In the latter half of the nineteenth century, when the idea of atoms became more and more convincing, Ludwig Boltzmann, an Austrian physicist, took the bold step of treating atoms as real objects that obeyed the same laws of physics that governed ordinary objects. In so doing, he departed from—and was subjected to harsh criticism, even outright ridicule (not least of which was due to his diminutive height), by—his fellow physicists who thought of atoms merely as a hypothetical convenience that was helpful in explaining certain physical phenomena.

What intensified his opponents' criticism was Boltzmann's injection of probability in the study of many-body systems. He argued that the number of atoms in a typical object is so large that keeping track of the behavior of each atom is impossible. Instead, one has to apply the laws of probability to this large number of atoms, killing two birds with one stone: on the one hand, the large number of constituents makes the system a great candidate for probability theory, which demands a large sample; on the other hand, it relaxes the impossible task of keeping track of the behavior of an unbearable number of atoms individually.

Boltzmann's fellow physicists, having had serious difficulty accepting the reality of atoms, could not bear the added headache of applying probability to their behavior. After all, was physics not an exact science? The opposition of his colleagues and the sense of loneliness caused by it may have had a role in the deterioration of Boltzmann's mental health and his eventual suicide in 1906, only one year after Einstein—employing Boltzmann's probabilistic methodology—proved the existence of atoms by explaining the so-called Brownian motion.

The recognition of Boltzmann's ideas predated Einstein's work when, in 1900, Max Planck—the German physicist who discovered quanta of radiation and who was one of the initial opponents of probabilistic atoms—in an attempt to make sense of his discovery of a certain formula that accurately described why hot objects glow the way they do, found no alternative but to accept Boltzmann's approach. Only then could Planck discover the quanta of radiation. Planck recognizes Boltzmann in his Nobel speech:

This problem ... [did not lead] me automatically to a consideration of the connection between entropy and probability, that is, to Boltzmann's trends of ideas, until after some weeks of the most strenuous work of my life, light came into the darkness and a new, hitherto undreamt of perspective began to open up for me.[7]

Boltzmann's work turned thermodynamics into *statistical mechanics*, and in the process made sense of some of the puzzling concepts in the field. For example, the concept of entropy entering in the second law of thermodynamics was a mystery, with which physicists knew how to deal mathematically, but could not comprehend physically. Why does entropy *always* increase in thermodynamic processes as the second law states? There was no satisfactory answer. However, once Boltzmann defined entropy in terms of probability, things started to make sense. Just like the painted coins in our salt shaker which *always* tend to the equality of black and white (perfect gray), or equivalently, to the maximum of probability, so do thermodynamic processes tend to maximize probability (or entropy). Why does heat go from a hot object to a cold object—and not in reverse—when the two come in contact with each other? Because the first way increases the entropy (maximizes probability) while the second way decreases it. Why does the smoke of a pipe rise, and once risen, never comes back to the pipe? Because the first process maximizes the probability, and the second process decreases it. Why do cars have to have exhaust pipes and factories chimneys? Because the conversion of the energy in the fuel (gasoline or coal) entirely into mechanical energy decreases the probability, and in order to increase it, some of that energy must be dumped into the environment.

That the second law of thermodynamics is given in terms of probabilities may seem paradoxical: laws of physics are exact; probability is anything but. However, as we learned in this chapter, a collection of a huge number of microscopic particles, even in a seemingly small thermodynamic system, behaves in an exact manner even though its constituents behave chaotically.[8]

Quantum Tunneling

Boltzmann applied the theory of probability to the entire collection of atoms. In his treatment, atoms obeyed the classical *deterministic* laws of physics. His assumption was that, if you could keep track of a single atom, you would see that it moves with the same exactness that a planet moves around the sun: you could *determine* the motion of a single atom exactly, once you knew the forces causing that motion. Boltzmann's theory eventually made sense to physicists, even though it encountered backlash from the physics community at first. As long as the behavior of each individual atom remained deterministic, physicists could accept the probabilistic theory of a large aggregate of atoms.

Erwin Schrödinger introduced his eponymous equation, which described the behavior of a *single* electron in a hydrogen atom, in January of 1926. It was considered a wave equation with a function that contained information about the behavior of the *wave* associated with the electron, which was known to be a *particle*. So, there emerged a puzzle: Is electron a wave, or a particle, or both?

After about six months of debate and speculation, Max Born suggested that the electron is a particle and the function in the Schrödinger equation is the *probability* amplitude, the square of whose absolute value gives the probability of the behavior of the electron in a hydrogen atom. This idea departed from Boltzmann's: not only does a collection of hydrogen atoms behave randomly, but, contrary to the deterministic laws of classical mechanics, each individual atom does the same. This was too much for the physics community to handle. Even Einstein, who embraced Boltzmann's probability and contributed to it, could not accept chaos at this fundamental level and blasted Born by famously saying "God doesn't play dice." … But, despite Einstein, He does! To see Him play dice, imagine we could actually observe quantum phenomena directly. Imagine we were "quantum" beings.

Before shrinking to a quantum size, let's put an ordinary candy in an ordinary jar, and place the jar in a cupboard. Go and check on it an hour (or a day, a week, a month) later. In the absence of a mischievous child (or a weak-willed adult), the candy will be in the jar. Now flip the "shrink" switch and turn into a quantum person. Then take a quantum candy and store it in a quantum jar. What should you expect in this quantum jar-candy system?[9]

There is a Schrödinger equation for such a system, which can be solved, and the solution predicts the probabilistic behavior of the candy. Question: Is it possible for the candy to be found outside the jar? Answer: Yes, there is a nonzero probability that the candy will find its way out! Physicists, for lack of a better word, call this weird process *tunneling*. But there is no tunnel in the wall of the jar. It is as solid as they get. Nevertheless, one moment the candy is inside, and next moment it is outside. Like magic! You can tape the process with a video recorder; all you'll see is that in one frame the candy is inside and in the next frame it is outside. No path, no tunnel, no recording of how it got there. As strange as this sounds, quantum tunneling explained the hitherto unexplained (alpha) radioactivity and fusion of nuclei. It also prompted the invention of *tunnel diodes*. But the biggest surprise came in 1981.

A *scanning tunneling microscope* (STM) has an extremely sharp conducting tip. When the tip is brought very near to the surface of a sample, electrons from that sample "tunnel" through the vacuum and end up on the tip. This causes a "tunneling current," which can then be electronically transformed into an image on a computer screen. The magnification of STM is so large that it can see atoms ... literally! In 1981, for the first time in history, humankind got a glimpse of what Greek atomists conjectured almost twenty-five centuries earlier.

Once physicists could *see* atoms, they set out to *manipulate* them ... move them around. First, they moved them to spell a famous acronym,* then to create the smallest movie ever.† Aside from fun, *nanotechnology* has become the new frontier of inventions with far-reaching applications in medicine, information technology, transportation, energy, food safety, and environmental science, among many others. And it is all thanks to (probabilistic) quantum theory.

Non-locality

Extrasensory perception (ESP) is the faculty of perceiving things by means other than the known senses, e.g., by telepathy or clairvoyance. ESP, telepathy, and clairvoyance are paranormal phenomena catered by mystics and "studied" by parapsychologists. The prefix "extra" meaning outside or beyond, segregates the senses from what is allegedly perceived.

Modern physics—relativity and quantum theories—is, one might say, "extra" sensory. Our senses are incapable of detecting phenomena in the domain of modern physics in which objects are too small or move too fast or both. As an example, consider the relativistic slowing of time: if an astronaut moves to a distant star and returns with a speed nearly equal to the speed of light, they may be in their forties while their surviving fellow astronauts left behind, who were the same age before the journey, may be in their eighties. Can we "see" such time warps? In a typical round trip to the moon, the difference in aging is about 0.00003 second! We can't "sense" that slight difference. Relativity is "extra" sensory for us, and this can motivate some to mystify relativity, as did Eddington in 1920.

A mystical relativity did not catch on among the public, partly because the founder of relativity, Einstein, was a fierce and vocal opponent of mysticism. The founders of quantum physics, on the other hand, were not only mystics

---

*https://www.sciencephoto.com/media/775020/view.
†https://www.youtube.com/watch?v=oSCX78-8-q0

themselves, but attached their mysticism to their science, as the narrative of the last chapter made clear. Furthermore, certain quantum physical puzzles defied explanation for a long time, which gave mystics opportunity to further their mysticism. An example is the EPR paradox.

Albert Einstein and two of his colleagues, Boris Podolsky and Nathan Rosen, wrote a paper in 1935 in which they tried to show that quantum mechanics was not a complete theory and needed certain *hidden variables* to make it complete. At the heart of their argument was the fact that, based on the principles of quantum physics, the measurement of a property of one particle at one location can affect the measurement of the same property of a second particle—which was the "partner" of the first one at the moment of creation of the pair—even though the two measurements are performed in laboratories hundreds of miles apart. Einstein declared this "spooky action at a distance" preposterous. He suggested that some hidden variables were needed to cure the disease. Later, he deemphasized the hidden variables and focused on the non-locality as a blemish on quantum physics.

It took almost thirty years to resolve the issue. In 1964, John Bell invented his own version of the EPR paradox. He assumed the existence of some very general hidden variables and, based on that assumption, derived an eponymous mathematical inequality using those variables, and showed that the prediction of quantum physics is not compatible with *Bell's inequality*. If quantum physics is correct, then no hidden variable theory can rescue us from the non-locality Einstein considered so preposterous. Furthermore, Bell's inequality suggested an experiment to settle the question of non-locality once and for all. In 1980, Aspect, Granger, and Roger reported an experiment which was in excellent agreement with quantum physics and clearly incompatible with Bell's inequality. *Nature is non-local!*

If quantum physics says that nature is non-local, then what's wrong with claiming that it is possible to move objects at a distance by the power of the mind … (Extrasensory perception, telepathy, clairvoyance)? There are two things wrong with the claim. One, which the mystics refuse to accept, is that the strangeness of quantum physics is confined to the microscopic world: the world of atoms and subatomic particles. The world of our senses is very much local. The second is confusing locality with causality. Nature is non-local, but it is not non-causal. You cannot physically affect the state of an object out there (target) without sending another object (probe) that is capable of changing that state. In Aspect, Granger, and Roger experiment, there is no control over what happens at B once the measurement is done at A. And since nothing can travel faster than light, cause and effect are always separated by time, unlike the Aspect, Granger, and Roger experiment, in which the influence at B occurs simultaneously with the measurement at A.

We are approaching the one-hundredth anniversary of the publication of Schrödinger's equation for an electron in a hydrogen atom and Born's probabilistic interpretation of the function in that equation. In the intervening years, the equation and the probability attached to it have been applied to other atoms, molecules, solids, liquids; generalized to include relativistic effects and the combination applied to the tiniest particles and largest cosmic objects, including the universe itself, and each time the theory has been remarkably successful. We owe the invention of lasers, transistors, light emitting diodes (LED light bulbs), microchips, superconductors, and scanning tunneling microscopes to the combination of Schrödinger's equation and Born's probabilistic interpretation of it.

Quantum physics is a complete theory and if you talk to it—to paraphrase Galileo—in "the language of nature," i.e., mathematics, it is consistent and makes perfect sense. If it behaves strangely, it is because of the prejudices we accumulate over time due to the incompetence of our senses and the limitation of the brain of each of us individually. In some sense, it is us who are strange. After all, what is more natural and non-strange that Nature itself?

Quantum physics is *not applicable to a macroscopic system* as a whole: there is no Schrödinger equation for solar system, an airplane, a truck, a racing car, or two colliding billiard balls. Macroscopic objects, as a whole, obey *deterministic* Newtonian physics. The purveyors of New Age mysticism, however, want you to believe that the macroscopic world is haphazard because the most fundamental theory of physics is probabilistic; that cause and effect are intertwined; and that the behavior of Brahman and Atman of Hinduism is aligned with quantum physics.

We can never "explain" quantum tunneling, the uncertainty principle, or non-locality using our mother tongue. Nevertheless, we can appreciate the weirdness associated with quantum physics' probabilistic nature by examining the weirdness of familiar random events such as the tossing of coins, as we did earlier in this chapter. There is no analogue for non-locality in the macroscopic world. But just as associating consciousness to a quantum particle based on its probabilistic nature would imply a manifestly nonsensical idea of a conscious coin, could we not infer that non-locality does not validate consciousness, universal soul, telepathy, and clairvoyance?

5

From Duality to Mysticism

Quantum physics came into being as a result of the investigation of the nature of the emission of light—more accurately, electromagnetic waves (EMWs)—by hot objects; specifically, the relation between the color of light emitted by an object, say a metal, and the temperature of that object. As generations of blacksmiths have witnessed, iron glows as you heat it. The glow starts with a dull red color, which changes to bright red as you keep it in a hot furnace, and finally to a white glow if the furnace is sufficiently hot. What they didn't know is that the iron glows EMWs *all the time*, even at low temperatures. However, due to the invisibility of EMWs that lie outside the boundaries of rainbow colors—red, orange, yellow, green, blue, indigo, and violet (the range of EMWs to which our optical nerves are sensitive)—they were not aware of that fact.

Once the Scottish physicist, James Clerk Maxwell, unified electricity and magnetism in 1865 and—in what could arguably be called the most important event of the nineteenth century—predicted the existence of EMWs, interest in their relation to heat exploded. After a frenzy of activities in laboratories and theoretical research in the last two years of nineteenth century aimed at unraveling the heat-EMW alliance, Max Planck explained the phenomenon by assuming that EMWs emanating from a hot object consisted of "bundles of energy" called *quanta* and that the energy of each quantum was proportional to its frequency. The constant of proportionality is now called the *Planck constant*, an extraordinarily small quantity, which, in scientific units, is 0.000 ... 00626 (replace the dots with 28 more zeros).

A related unexplained phenomenon was the *photoelectric effect*, whereby certain metals produce electric currents when light of sufficiently high frequency is shined on them. In 1905, Einstein explained the effect by assuming that all EMWs, not just those emitted by hot objects, consisted of actual particles—rather than bundles of energy—which he named *photons*, whose energies were proportional to their frequencies a la Planck.[1]

Yet another puzzle was the structure of atoms. In 1897, when he discovered electrons, J. J. Thomson proposed the *plum-pudding* model of the atom. Based on this model, the negative electrons (the plums) are embedded evenly in the positive background charge (the pudding). Ernest Rutherford set out to test Thomson's model experimentally by bombarding gold atoms with alpha particles, one of the three unusually energetic "rays" discovered at the turn of the last century christened after the first three letters of the Greek alphabet. To his surprise, Rutherford discovered, in 1911, that the positive charge of the gold atom—and, by generalization, of all atoms—was not like pudding, but was highly concentrated and extremely heavy. He called it the *nucleus* of the atom.

Now the atom looked like a miniature solar system with the nucleus being the sun and the electrons moving around it like planets. However, this *planetary model* of the atom was dead on arrival, because, unlike planets, the electrons are electrically charged, and charged particles in circular motion emit EMWs, which carry energy. As the electrons donate their energy to the EMWs, they cannot maintain their orbits and spiral into the nucleus, ending the life of the atom.

A Gift from Nature

Hydrogen is called "the gift of Nature." With just one electron and a nucleus, it is the simplest atom in existence and, therefore, the most ideal element for theoretical investigation. Deciphering hydrogen was a roadmap to understanding the other complicated atoms. How does the electron move around its nucleus in the H-atom? Niels Bohr took the bold step in 1913 by theorizing that the Planck constant is an elemental angular momentum and that the angular momentum[2] of the electron, as it circles the nucleus, is quantized,* meaning that it is an integer multiple of the Planck constant.

By combining the *classical* equation of motion of the electron with his *quantum* hypothesis, Bohr calculated the radii of orbits of the electron and the energy associated with each orbit and proved that the radii and energies were

---

*Something is said to be quantized when any of its values is an integer multiple of an elemental unit.

both quantized and the larger the radius of an orbit, the larger the energy of that orbit.[3] In this *Bohr model* of the H-atom, the electron can be only on those allowed quantized orbits, the smallest one of which—called the *Bohr orbit*—keeps the electron at a minimum distance from the nucleus and prevents it from collapsing into the nucleus. If the electron happens to be on a large orbit, it can "jump" to any one of the smaller orbits and emit a photon whose energy is the difference between the energies of the two orbits.

The unique energy of the photon gives it a unique color (more precisely, frequency or wavelength)[4] determined by Planck-Einstein relation. Different orbit-to-orbit transitions give different photon wavelengths, which are also quantized. The collection of these wavelengths is called the *spectral lines* of the H-atom.[5] The spectral lines of the H-atom had been observed before Bohr, but there was no theoretical explanation for them. Bohr's theory, for which he won the 1922 Nobel Prize in Physics, explained the spectral lines of the H-atom.

As good as his model was, Bohr was not satisfied with it because of the arbitrariness of his underlying assumption.[6] A partial lift to this arbitrariness was provided in 1923 by Louis de Broglie, a French physicist. De Broglie took the reverse step of Einstein and Planck: if electromagnetic waves are made up of *photons*, then maybe electrons are the particles of some kind of a *wave*. He proposed that the wavelength (the wave property) and momentum (the particle property) of an electron are related by the same formula that connects the momentum and wavelength of a photon,[7] namely that wavelength is Planck constant divided by momentum. This *de Broglie* relation explained why Bohr's starting assumption concerning the angular momentum of the electron in a H-atom made sense.

If electron is a wave, how can we demonstrate its "waviness"? More generally, how does one establish the wave nature of a physical entity? When waves are created by two appropriately prepared sources, they create an *interference pattern*.* The pattern is created because at some points, the crest of the wave of one source meets the trough of the wave of the other source and cancels it (destructive interference). At other points two crests or two troughs meet and reinforce one another (constructive interference). At destructive interference points no wave exists, while at points of constructive interference multiply intensified waves are created.

---

*https://www.youtube.com/watch?v=kPe1z4GC1uU shows a nice demonstration of the interference of water waves.

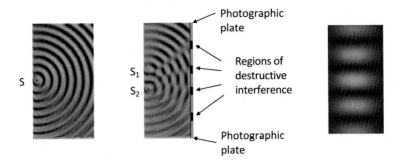

Fig. 5.1 Interference pattern created by the two sources S_1 and S_2

In Fig. 5.1, each of the two sources in the middle (S_1 or S_2) produces a wave like the source S on the left. Together they create the interference pattern shown. For subatomic particles, the two sources could be two closely spaced slits—a *double-slit*—cut out in one screen through which the particles pass to be later collected on a second detecting screen such as a photographic plate for photons. Interference pattern is bands of bright (corresponding to constructive interference) and dark (corresponding to destructive interference) lines on the plate. In the case of light, the radially outward regions of destructive interference produce dark areas on the photographic plate when the waves arrive there as shown in the plot on the far right of the figure. The detection of the interference pattern for light led Thomas Young to conclude in 1803 that light was a wave.

From Uncertainty to Complementarity

The interference pattern of the electrons was demonstrated in two simultaneous experiments done in 1927, confirming de Broglie's hypothesis. Generalizing from photons and electrons, physicists concluded that microscopic physical entities are both particles and waves. This came to be known as the *wave-particle duality* and is at the heart of the mysticism imposed on quantum physics. It appeared that at a fundamental level, reality had a dissociative identity disorder. Always fond of philosophizing, Bohr advanced this apparent duality to its full philosophical and mystical potential.

In February 1927, Werner Heisenberg published his celebrated *uncertainty principle*, which states that the product of the uncertainties in position and momentum of a particle is larger than the Planck constant. Uncertainty is understood in terms of the statistical spread—also called *standard deviation* in statistics and probability—in values obtained when a quantity is measured many times. As an illustration, suppose you fire a bullet from a gun held at a particular vertical angle. Two seconds later, measure the location and momentum of the bullet with the most accurate devices available. Fire the bullet again and measure its location and momentum as before. In general, the two measurements differ—ever so slightly—because of the limitation of the accuracy with which they are made.[8] Repeat the experiment a thousand times—or, equivalently, fire a thousand identical guns at the same time—and calculate the uncertainty (spread, standard deviation) in the thousand values of position and momentum. Heisenberg's uncertainty principle says that the product of the two uncertainties must be larger than the Planck constant. For bullets, this poses no restriction on the accuracy of measurements because the two uncertainties are so large that their products necessarily is much larger than the Planck constant.[9] Therefore, we can say with utmost confidence that *the position and momentum of large objects can be measured simultaneously with arbitrary accuracy*, in agreement with classical physics.

The last sentence captures the essence of the irrelevance of quantum physics to everyday phenomena.[10] If you want to see whether quantum physics is relevant to a phenomenon, apply the uncertainty principle to it—when applicable. If Planck constant is too small compared to the quantities measured in the phenomenon, quantum physics is irrelevant. For all everyday phenomena, quantum physics is irrelevant. Insisting on its relevance is in the playbook of mystics who want to inject the weirdness of quantum physics into the world that surrounds us.

The situation is different for an electron or any other subatomic particle. The values obtained for position and momentum of an electron are so small—and therefore the uncertainties in them so much smaller—that when you multiply the two you get a number that is comparable to Planck constant. Heisenberg's uncertainty principle puts a limit on how small the product can be. We cannot measure the momentum and position of a subatomic particle with arbitrary accuracy. The more certain we are of the position of a subatomic particle (i.e., the smaller the uncertainty in position), the less certain we will be about its momentum (i.e., the larger the uncertainty in momentum). When we are absolutely certain about the position of a particle, i.e., when the uncertainty in position is zero, we are completely ignorant about its momentum, i.e., the uncertainty in momentum is infinite.

Bohr had created an institute at the University of Copenhagen—now called the Niels Bohr Institute—in which the inventors of the new quantum physics gathered and collaborated. In this institute, Bohr and Heisenberg discussed the quantum physics intensely, sometimes for long hours into the night. Bohr, who always had a keen interest in philosophy, tried to "make sense" of Heisenberg's uncertainty principle in philosophical terms. Momentum is related to the wave property via the de Broglie relation. Position, on the other hand, presupposes localization, which is a particle property.

To Bohr, the uncertainty principle turned into a statement about the wave and particle nature of a subatomic entity: if we are certain that a physical entity is a wave, then its particle property cannot be detected and vice versa. He thus transformed the mathematically precise uncertainty principle into the philosophically vague statement that the wave and particle nature of an electron complement each other but are mutually exclusive. This has come to be known as the *complementarity principle*.

According to this principle, if you set up an experiment that measures the particle nature of an electron, then its wave nature evaporates. On the other hand, if an experiment detects the wave nature of the electron, its particle nature evaporates. The experimenter decides whether an electron is a wave or a particle. *The observer creates the reality of an electron*. This strikingly echoes Schopenhauer's description of reality:

> It is the human being that, in its very effort to know anything, objectifies an appearance for itself that involves the fragmentation of Will and its breakup into a comprehensible set of individuals.

The notion of an observer-created reality implied by the complementarity principle throws mystics into a fit of ecstasy because they now can claim that the old scriptures like *the Upanishads* foretold modern physics and that the old wisdom is based on science. When this conclusion is fortified by politics, especially one with potent nationalistic overtone, it becomes monstrously bizarre.

The 106th Indian Science Congress, which was inaugurated by Prime Minister Narendra Modi, ran from 3 to 7 January, 2019. The head of a southern Indian university cited an old Hindu text as proof that stem cell research was discovered in India thousands of years ago. G. Nageshwar Rao, vice chancellor of Andhra University, said a demon king from the Hindu religious epic, Ramayana, had 24 types of aircraft and a network of landing strips in modern day Sri Lanka. Another scientist from a university in the southern state of Tamilnadu told conference attendees that Isaac Newton and Albert Einstein were both wrong and that gravitational waves should be renamed "Narendra Modi Waves."[11]

The Double-Slit Experiment

Observer-created reality is best understood—and debunked—by the analysis of the *double-slit experiment* for subatomic particles such as electrons or photons. To appreciate the oddity of the experiment, first consider a similar large-scale experiment whose outcome we can anticipate. Let a gun fire bullets toward a vertical barrier with two identical vertical rectangular slits, which allow bullets to go to the other side.[12] The bullets are subsequently collected on a second barrier made of cork. We perform three different experiments with this gun-barrier set up.

In the first experiment, we cover one of the slits and fire a large number, say 1000, bullets. After all the bullets are fired, we go to the other side and count the number of bullets collected by the cork. Let us say that 50 bullets made it to the cork barrier. It then follows that the probability—call it P_1—for the bullets fired from the gun to pass through the first slit is 50/1000 or 0.05 or 5%. The 50 bullets that passed through the slit will form a vertical blob. In the next experiment, we cover the other slit—while uncovering the first—and fire 1000 bullets as in the first experiment. Since the two slits are assumed to be identical, we expect that around 50 bullets will make it to the other side. Thus, the probability—call it P_2 this time—for the bullets to pass through the second slit is also 0.05 or 5%. The second blob will look more or less like the first. Finally, in the third experiment, we leave both slits open and fire 1000 bullets as in the first two experiments. If we count the number of bullets collected by the cork, we will see that around 100 bullets will have made it to the other side. Thus, the probability—call it P_{12}—for the bullets to pass through both slits is 0.1 or 10%. This probability, as expected, is simply the sum of the other two probabilities, and the bullets making it to the collector made of cork form two identical (possibly overlapping) vertical blobs.

Photons can be used as microscopic bullets in the double-slit experiment. Take a solid screen and cut two identical tiny slits in it that are very close to one another. Now fire photons one by one from a "photon gun" toward the two slits. Put a photographic plate behind the first screen to pick up the photons that pass through the two slits. Will we see an outcome similar to the bullet experiment? After all, photons are just tiny bullets. Or are they?

When each slit is open by itself, a blob appears on the photographic plate analogous to the pile of bullets collected by the cork in the previous experiment.[13] Now, we open both slits and wait until a sufficiently large number of photons have passed through them. What will we discover? Will we see two blobs as in the gun-bullet experiment? Surprisingly, no! Instead of two blobs, we see a pattern of bright and dark fringes. This is indeed baffling! And when interpreted outlandishly, it can be exploited to impart consciousness to photons.[14]

Explaining the Double-Slit Experiment

The proper way to explain the outcome of the double-slit experiment for photons is to employ the Schrödinger equation, which contains the ubiquitous symbol Ψ (pronounced "sigh"). The symbol is called the *wave function* or *probability amplitude*.[15] Ψ has two fundamental properties: (1) the square of its absolute value is the probability of the behavior of the system it describes, and (2) it obeys the *superposition principle*: if there are several paths available to the system, the total Ψ is the sum of the Ψs for each path.

In the case of the double-slit experiment, when only the first slit is open, the wave function is, say Ψ_1, and its square gives the probability function (or distribution) for the photon to go through the slit. This probability function determines the shape of the image of the slit on the photographic plate when a very large number of photons pass through it. The image turns out to be a blob on the plate. A time exposure reveals the individual photons arriving at the plate and landing somewhere in the blob completely randomly, with a probability that is larger at the center of the blob: the photons are more likely to land in the middle of the blob than at its rim.[16] As the exposure time increases, the final picture of the slit is revealed.

When only the second slit is open, the wave function is, say Ψ_2, and its square gives the probability for the photon to go through the second slit. Because the two slits are identical, the second probability is the same as the first, meaning that the shape of the second blob on the plate is identical to the first.

Now we open both slits, expecting to see just two blobs as in the experiment with bullets. But that is not what we get. Instead, we observe alternate bands of light and darkness. How can that be? The answer is simple. Since photons have two paths to take, the total probability *amplitude* for them to be detected at the photographic plate—by the superposition principle—is simply the sum of the two amplitudes, $\Psi_1 + \Psi_2$. To find the total probability, we must square this total amplitude; and it is well known that the square of a sum is not equal to the sum of the squares: There is an extra term, and that term is responsible for the bright and dark bands on the photographic plate.[17]

In a time exposure for the third experiment, the arrival of the photons on the photographic plate is demonstrated to be completely random, with

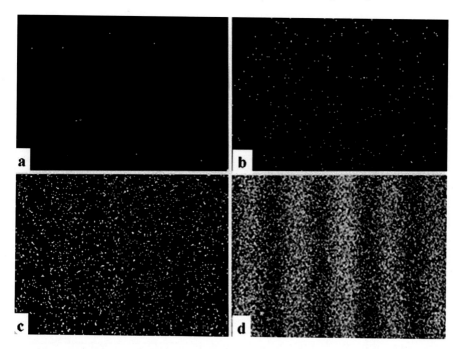

Fig. 5.2 The random distribution of individual electrons collectively yield a wave interference pattern

a probability that varies across the plate. In certain regions of the plate this probability is zero, that is, no photon is expected to land there. Between two adjacent dark regions, the probability is large, meaning that a large number of photons are expected to land there. This explains the alternate bright and dark regions on the plate when both slits are open. Figure 5.2 shows the build-up of *electrons* for the double-slit experiment. Part (a) of the figure depicts a few random dots, corresponding to the first few electrons that have arrived at the plate. In (b) you see more random dots, but still with no detectable pattern. It is with (c) that a pattern seems to be vaguely emerging. Once the plate is exposed for a long time, the complete grainy bright-dark pattern appears as in (d).*

The superposition principle and the probability interpretation of Ψ is the only way we can analyze the double-slit experiment. They explain the experiment whether you send an army of photons to the slits at once, or fire them one by one. It is only when we try to "understand" or "make sense" of the

---

*https://bit.ly/3MRWfeq shows an animated version of the figure.

double-slit experiment in terms of our day-to-day experience that we end up in a consciousness pitfall. And this pitfall is not restricted to quantum physics: any attempt at explaining any probabilistic outcome leads to nonsense as we saw in Chap. 4. Richard Feynman cautioned his students that if they tried to understand quantum physics, they would go down a drain from which no one has escaped.[18]

The Myth of Duality and Complementarity

The double-slit experiment can shed some light on the ostensible veracity of the complementarity principle as well as its falsehood. First, I'll demonstrate the pitfall in which ordinary language can throw us. Send a beam of light to the two open slits. An interference pattern will appear on the photographic plate, showing the wave nature of light. We also know that the beam consists of photons, each of which must go through either one or the other slit. Those photons that go through slit 1 must create a blob on the photographic plate. Those that go through slit 2 create a second blob. So, the overall pattern should be two—possibly overlapping—blobs, not an interference pattern! What is going on?

The answer is that the assumption "a given photon must go through either slit 1 or slit 2" is a lingual statement that does not make sense in quantum physics. It relies on our intuitive deterministic picture of a particle with a path (a line), which connects a point of the source to a point on one of the slits and finally to a point on the photographic plate. The assumption does not hold in probabilistic quantum physics where paths do not exist because they correspond—at a particular time—to an exact value of the location of the particle, and thus to a zero uncertainty in position, which is a violation of the uncertainty principle.

What if we could *actually* label the photons that go through either slit? Then we would have a way of knowing which photon went through which slit. Would we see two blobs on the back screen? If so, then the complementary principle—and its extension, observer-created reality—ostensibly holds: if we engage the labeling device, and thus determine the "path" of the photon, we make it a particle; if we don't engage the labeling device, we turn the photon into a wave. We create the reality of the electron! No, we don't! Engaging or not engaging a device is not creating the reality of the photon. It is changing the experiment altogether. And we should expect different results for different experiments … even in classical physics. Let's continue with the experiment involving labels and analyze the disappearance of interference.

It so happens that there *is* a way of labeling photons. Light has a property called *polarization* that can be thought of as an arrow pointing in some direction. We can speak of a light having a vertical polarization or a horizontal polarization. If you send an unpolarized beam of light through a *polarizer*, the emerging light will have a polarization determined by the direction of that polarizer: if you hold the polarizer vertically, the emerging light will be vertically polarized, and if you hold the polarizer horizontally, the emerging light will have a horizontal polarization. A beam of light that is vertically polarized cannot pass through a horizontal polarizer and vice versa.

Cover slit 1 with a horizontal polarizer and slit 2 with a vertical one and send a beam of unpolarized light to the double-slit. For simplicity, let's assume that the unpolarized light consists of photons that are either vertically or horizontally polarized.[19] Now, upon detection, we *know* which photons went through which slit. If we see that a photon is horizontally polarized, it must have come from slit 1. If it is vertically polarized, it must have come from slit 2. We have determined that the beam of light consists of particles and also which slit each particle went through. What is the pattern on the photographic plate? It is two blobs! No interference, and therefore, no wave property. Without polarizers, we couldn't tell which slit each photon went through, and therefore, whether the beam consisted of particles; so, it behaved like a wave. With polarizers, we *could* tell that the beam consisted of particles because we knew which slit each particle went through; so, its wave property evaporated. The Complementarity principle seems to hold. Have we created the reality of the photon?

The interference pattern does not really prove the wave nature of light. It only *confirms the probability distribution* predicted by quantum physics for the behavior of a collection of *particles*. Without polarizers, the probability function predicts a pattern that resembles wave-like interference; with polarizers, the probability function predicts *absence* of interference.

To elucidate the last statement, let's repeat the two experiments using single photons, first without polarizers. Suppose that instead of a photographic plate, we have a photon detector that "clicks." Send photons one by one in succession to the slits until you hear a click. Record the position of the photon on the clicking detector. *There is no wave anywhere*! Only a single particle triggering a clicker. Now cover the slits with polarizers and send photons one by one in succession until you hear a click. Once again, record the position of the photon. Except for the location of the photon—which is determined by the probability function in each experiment—there is no difference between the two experiments. In fact, even the two locations may be the same, because there is an overlap between the two probability functions.

There is no wave-particle duality; photon is always a particle. Since there is no wave, the complementarity principle does not make sense. Of course, if you keep sending photons so that a large number of them make it to the other side, the shapes of the two probability functions emerge: the function in the first experiment resembles a wave interference pattern, while that of the second experiment lacks interference.

The real question is "Why placing perpendicular polarizers destroys interference?" Recall from the discussion of the double-slit experiment on page 56 that the probability amplitude for the photon to be detected at the photographic plate is the sum of the amplitude for going through slit 1 and the amplitude for going through slit 2. To find the total probability, we must square this sum; and when you square a sum of two terms, you get the sum of the squares of the two terms—which is the sum of the probabilities of going through the two slits—*plus* twice the "product" of the two terms.[20] This product is responsible for interference and depends on the detailed mathematical structure of the two amplitudes. It turns out that in the absence of polarizers, the product is nonzero, resulting in a probability function that resembles wave interference. Placing horizontal and vertical polarizers alters the two amplitudes in such a way that their product vanishes, resulting in a probability function with no interference term.

It is interesting to note that the "product" mentioned above—which determines the pattern of the probability distribution—can be varied continuously from zero to a maximum. Install polarizers at the slits, whose polarizations are not necessarily perpendicular. By changing the relative orientation of the polarizers, we can change the likelihood that a photon with polarization corresponding to one slit appears to be coming from the other slit. This set-up gives us a "dial" by which we can control the behavior of many photons from a blob (zero interference) to maximum interference. Going clockwise from the top left to the bottom left of Fig. 5.3, the plots show the interference pattern detected when the angle between the polarizers changes from 90° (top left, when the "product" is zero), to 84°, to 73°, and finally to 0° (bottom left, when the "product" is maximum).

By changing the relative orientation of the polarizers, we are *not* creating the reality of a photon. We are simply preparing the state of a photon, thereby controlling the probability of where it lands on the detecting screen. Using a classical analogy, we are creating the reality of a photon as much as we are creating the reality of a bullet when we point our gun at a particular angle, thereby controlling the landing site of a projectile.

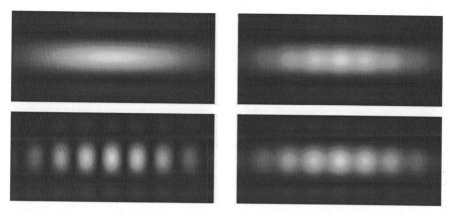

Fig. 5.3 As we change the angle between polarizers, the probability distribution of photons going through the double-slits changes

Objectivity Is Alive and Well

Bohr devised the complementarity principle in part as a seemingly scientific solace for his mysticism. He confided in Heisenberg:

> ... consciousness must be part of nature, or, more generally, of reality, which means that, quite apart from the laws of physics and chemistry, as laid down in quantum physics, we must also consider laws of quite a different kind. But even here I do not really know whether we need greater freedom than we already enjoy thanks to the concept of complementarity.[21]

One of the most pernicious consequences of Bohr's principle, or the wave-particle duality, is the attack on the very notion of the objectivity of science. If EMWs behave both as waves and as particles and the experimenter controls which one occurs, then reality is created by the experimenter and it is not objective. This conclusion is baseless as we saw earlier: particles have no wave property; the interference pattern is just the probabilistic distribution of a swarm of *particles*.

Another attack on objectivity exploits the probabilistic nature of photons directly. Two observers perform identical experiments in which a single photon is fired from a photon gun, passes through a double-slit, and lands on a detector plate. As the observers look at the locations of their photons on their plates, they note that they are not the same. Identical experiments lead to different outcomes. Do the experimenters influence the experiment? Are they creating the reality of the photon? Some mystics answer the question affirmatively and contend that the experimenter indeed influences the experiment. Other mystics impart intelligence to photons: the process is random; therefore, photons must make a choice, and only intelligent entities can make choices.

The death of objectivity entails the unity of subject and object and the entanglement of mind and matter. In a world in which the observer determines the outcome of an experiment, and therefore creates reality, mind rules over matter. This proposition is the tenet of Eastern mysticism and a golden gift to pseudoscientists who—as they inculcate it in the consciousness of the public—claim that it is as solid as quantum physics itself.

Objectivity is also attacked by defining it as the existence of objects irrespective of whether we observe them or not. Sitting in your room at night reading this page, you know that there is a Moon in the sky, possibly hiding behind a cloud. Whether you go out and look at it or not is irrelevant. A bullet approaching a blocking screen with two slits will go to the other side by passing through either slit 1 or slit 2. We don't have to set up detecting devices at the slits to verify that. We know that if we insert a thermometer in boiling water at sea level, it will read 100 °C. We don't need to actually read the thermometer to create the reality that water boils at 100 °C at sea level. It is in this context of objectivity that the proposition "an electron, being definitely a particle, has to go through slit 1 or slit 2", becomes a legitimate statement, and the fact that it cannot be answered, argue the mystagogues, definitively points to the downfall of objectivity.

These examples of objectivity are really instances of *determinacy*. That the Moon is there whether we look at it or not presupposes that the Moon does not change its position randomly. That the bullet goes either through slit 1 or slit 2 is the consequence of the fact that its trajectory is not chaotic. That water boils at 100 °C at sea level is based on the multiply verified assumption that the boiling point of water is not random.[22]

We know that quantum physics is not deterministic; both the predictions and verification of those predictions rely on the laws of probability. Does probability imply lack of objectivity? When the outcomes of two identical experiments differ—due to the probabilistic nature of quantum physics—are we to conclude that objectivity is dead? What is objectivity in a world in which probability rules? The answer boils down to another question: How do we verify a mathematically precise probabilistic statement?

To verify the statement that the probability of getting two heads when tossing four coins is 0.375 (6/16 = 3/8, or three out of eight)[23] we have to toss four coins many many times. The larger the number of times we toss the four coins, the closer the fraction—of times two heads show up—gets to 0.375. When the number of trials approaches infinity, the likelihood that we get a fraction *different* from 0.375 approaches zero.[24] This is the essence of probability theory and statistics, and the reason why in actuarial and polling applications, very large samples are desired, knowing that small samples are not reliable as predictors of the behavior of the population. That is also why we can talk about the temperature of a substance as if it could be measured exactly. Even though temperature is a statistical quantity related to the average kinetic energy of the particles making up that substance, there are practically infinite number of those particles even in a small volume of the substance.

If objectivity is the absence of the influence of the subject on the object, then in a world in which only the probability of occurrences can be predicted and observed, any act of measurement has to consist of a very large sample.

> If one observer repeats a given event many many times to verify the predicted probability of the event, and a second observer does the same, and the two observe the same (actually very *nearly* the same) outcome, we are justified – in fact, we are required – to call the result of the observation independent of the observer.

This independence of the observation of a probabilistic event from the observer is the only sensible definition of objectivity. That two observers get two different results when performing an experiment involving a single microscopic event—or a small number of events—does not overthrow scientific objectivity, nor does it prove that the experimenters somehow influence the experiment.

Almost a century has passed since the philosophical and political turmoil in which quantum physics was born. As practicing physicists are exploring the frontiers of the vast scientific territory that quantum physics opened to us, the philosophical scar inflicted by its founders is not going away. The same quantum physics that has given us lasers, light emitting diodes, microchips, and scanning tunneling microscopes, and has unraveled the secrets of the tiniest constituents of matter and the largest galaxies hitherto unseen, that same quantum physics has become an effective tool for thrusting nonsense into the collective mind of the public. New Age gurus regurgitate mystical quotes by Bohr, Heisenberg, Schrödinger, and Pauli to convince the unsuspecting public that, based on the exact science of quantum physics, reality is in our mind and, therefore, simply thinking about health, happiness, and wealth makes them happen. If the mysticism had stopped with the founders, the sheer passage of time might have dissociated quantum physics from spirituality. But some later influential physicists had too strong an interest in philosophy to let that happen.

6

Quantum Consciousness Crosses the Atlantic

In many respects, physics at Princeton in the 1950s resembled physics at Copenhagen in the 1920s. In both cities, great scientific minds gathered together to solve outstanding problems of the time. If Copenhagen was the birthplace of quantum mechanics, then Princeton witnessed its adolescence in application and generalization. The similarity between the two cities goes farther. Philosophical speculations on matters beyond physics in Copenhagen, which contributed to the mysticism attached to quantum mechanics, continued in Princeton. J. Robert Oppenheimer, Freeman Dyson, John von Neumann, Eugene Wigner, and John Wheeler continued the philosophical and mystical tradition of Bohr, Heisenberg, Schrödinger, and Pauli.

Oppenheimer's Mystique

J. Robert Oppenheimer is best known for his role in the Manhattan Project. As the director of the secret weapons laboratory at Los Alamos, he was instrumental in developing the atomic bombs dropped on Hiroshima and Nagasaki. He was a brilliant physicist, having obtained his PhD at the age of 23 in 1927 under the mentorship of Max Born, one of the founders of quantum physics.

Oppenheimer once described himself as a properly educated aesthete. His colleague, Isidor Rabi was struck by how Oppenheimer, as a postdoctoral student in Zurich, conducted himself as a literary man rather than a physics student. Even at Los Alamos, Oppenheimer rarely talked about physics and weapons, instead, "he talked about the mystery of life. … He [would] walk around the room ... rub his palms together and look to the side ... He kept quoting Bhagavad Gita."[1]

The combination of scientific genius and poetic humanism gave Oppenheimer a mystique seen only in the most influential religious figures. Those in contact with Oppenheimer saw him as mesmerizing: hypnotic in private interaction, but often frigid in more public settings. At Berkeley, where he was teaching and doing research, his students were so impressed by his charisma that they often adopted his walk, speech, and other mannerisms.

After WWII, Oppenheimer left Berkeley to direct Institute for Advanced Study in Princeton, NJ. IAS was the powerhouse of scientific intellect to which were attracted many young talents, who were also awed by the presence of Einstein, a giant figure at the Institute since 1933. Under Oppenheimer's directorship, many bright physicists who shaped the future of physics in the US came to IAS either as visitors or as permanent faculty. Among such luminaries were Freeman Dyson (a mathematician and theoretical physicist who made significant contribution to both fields, including quantum electrodynamics); Murray Gell-Mann (a child prodigy who graduated from college at the age of 15 and received his PhD when only 21; he is most famous for the invention of the quark concept); T. D. Lee and C. N. Yang (two young Chinese physicists who discovered a unique and strange property of the weak nuclear force); and John von Neumann (another child prodigy who contributed to branches of mathematics as diverse as set theory and computational analysis).

Oppenheimer occupied a central place in the postwar physics community. "When a physicist came up with a new discovery, it was customary to make a pilgrimage to the Institute and try out the new idea on Oppenheimer and his young geniuses."[2] Judging from his charisma and communication skill, it is not unlikely that the new generation of physicists at IAS and Princeton University, which is at the proximity of IAS both geographically and academically, were exposed to Oppenheimer's theosophy and some were possibly influenced by it.

Dyson's Consciousness

Freeman Dyson, one of the recruits to IAS by Oppenheimer, remained at the Institute for more than half a century until his death in 2020. Dyson was a child prodigy—he calculated the number of atoms in the Sun at the age of five—and during his adult professional career, he made major contributions to both physics and mathematics. His most notable contribution to physics was in quantum electrodynamics where he proved the equivalence of two different,

6 Quantum Consciousness Crosses the Atlantic

highly mathematical, approaches to the interaction of light with electrons. The only thing that kept him away from the Nobel Prize was the Prize's restrictive rule that no more than three people can share it in the same category.

Dyson had also a speculative side. Some bold speculations drove Dyson to territories far beyond his areas of expertise and dangerously close to quack science. One of his speculations was to genetically engineer warm-blooded plants that could grow on comets. Speculation about future is not the only venue in which Dyson vents his ideas. He has thrown in theories on the origin and evolution of life, which disagree with the theories commonly held by evolutionary biologists and other experts. He doesn't mind creating an uproar about the climate change, a field in which he has no expertise and has done no research. And it doesn't bother him to admit that it is better to be wrong than to be vague![3]

Speculators cannot avoid philosophical, theological, and mystical questions, and Dyson is no exception. Although he warns that the design argument[4] is a theological and not a scientific one, and that it is a mistake to interject theology in science, Dyson strays away from his own warning and considers "the argument from design *to be valid* in the following sense."

> The universe shows evidence of the operations of mind on three levels. The first level is the level of elementary physical processes in quantum mechanics. ... The second level ... is the level of direct human experience. Our brains appear to be devices for the amplification of the mental component of the quantum choices made by molecules inside our heads. ... *Now comes the argument from design*. The argument here [for the evidence of the universe being hospitable to the growth of mind] is merely an extension of the Anthropic Principle up to a universal scale. Therefore it is reasonable to believe in the existence of a third level of mind, a mental component of the universe.[5] [Emphasis added]

Ideas such as those expressed in the quote above won Dyson the Templeton Prize in 2000. The Templeton Foundation, established by billionaire investor Sir John Templeton in 1972, gives out an annual Templeton Prize for "progress toward Research or Discoveries about Spiritual Realities," which has been designed to fill a gap left by the Nobel Prizes and pointedly pays more than they do. The Foundation's campaign to bring scientific legitimacy to religion has led to some dubious ventures, including funding in 1999 for a conference on Intelligent Design as an alternative to evolution. More recently, the foundation has backed away from intelligent design in favor of funding research into the efficacy of prayer in healing certain illnesses.[6]

The Anthropic Principle, to which Dyson refers in the preceding quote as—if it were—a scientific principle to strengthen his idea of a "third level of mind", is actually a theosophical idea injected into science by Dyson's fellow speculators. John Barrow and Frank Tipler wrote a book titled *The Anthropic Cosmological Principle*, in which they argue that the universe is the way it is because it was compelled to eventually have conscious and sapient life emerge within it. Tipler is also the author of *The Physics of Christianity*, in whose introduction he identifies the initial singularity—the big bang—as God. John Barrow is a cosmologist, mathematician, philosopher, and playwright. His philosophical writings, to which he adds the weight of his science, are tainted with the mystical notion of consciousness and what Dyson calls the "third level of mind." These philosophical writings pleased the Templeton Foundation to the point that it awarded Barrow the Templeton Prize in 2006.

Wigner's Consciousness

Fasori Evangélikus Gimnázium is a famous secondary school in Budapest, Hungary. In early twentieth century, it was one of the best schools in Budapest and was part of a brilliant education system designed for the elite. Even though the school was run by the Lutheran Church, it was populated predominantly by Jewish students, some of whom became world renowned scientists:

- Theodore von Kármán, a mathematician who is regarded as the outstanding aerodynamic theoretician of the twentieth century;
- George de Hevesy, Nobel Laureate in Chemistry, developer of radioactive tracers and co-discoverer of the element hafnium;
- Leó Szilárd, discoverer of the nuclear chain reaction;
- Dennis Gabor, electrical engineer and physicist, most notable for inventing holography, for which he was awarded the 1971 Nobel Prize in Physics;
- Edward Teller father of the hydrogen bomb; and
- Paul Erdös, the most prolific mathematician of the last century.

But the two most notable graduates of the school are John von Neumann and Eugene Wigner. John von Neumann was an incredible child prodigy. When he was six, he could divide two 8-digit numbers in his head and converse in Ancient Greek. By the age of 8, von Neumann was familiar with differential and integral calculus. At 15, he began to study advanced calculus under Gábor Szegö, one of the Hungarian leading mathematicians. By the time he was 19, von Neumann had already published two major mathematical papers. In

1933, he was offered a lifetime professorship on the faculty of the Institute for Advanced Study and remained there as a mathematics professor until his death in 1957.

Eugene Wigner was a year ahead of von Neumann at Fasori Evangélikus Gimnázium. When asked why the Hungary of his generation had produced so many geniuses, Wigner, who won the Nobel Prize in Physics in 1963, replied that von Neumann was the only genius. He himself was a giant in the world of physics and mathematics. Wigner's contributions to mathematical physics began during his studies in Berlin, where his supervisor suggested a problem dealing with the symmetry of atoms in a crystal. His friend John von Neumann pointed out the relevance of representation theory of groups, an active area of research in mathematics. Wigner soon became enamored with group theory and began to apply that approach to quantum mechanical problems. He lectured briefly at the University of Göttingen before moving to America to escape the Nazis. In 1930, Wigner accepted a visiting—which later turned into a permanent—professorship at Princeton University where, except for occasional visiting appointments in America and abroad, he remained until his death in 1995. Wigner's greatest accomplishment is his application of group theory to a combination of quantum physics and Einstein's special relativity and discovering the mathematical representation of a particle in 1939.

Wigner, like his West European contemporaries who discovered quantum physics, had a keen interest in philosophy, especially as it related to quantum mechanics. The interplay between philosophy and quantum wave-particle duality led Bohr, Heisenberg, Schrödinger, and Pauli to inject some elements of consciousness into quantum physics. Wigner, on the other hand, concentrated on the act of measurement and came up with a strong solipsistic version of consciousness. And he was helped by his friend and schoolmate, John von Neumann.

A quantum mechanical system, described by a certain wave function, is in a combined state of many potential outcomes (eigenvalues), each eigenvalue having its own probability. An apparatus designed to measure these eigenvalues can detect only one of them. A subsequent measurement of the system detects only the eigenvalue already measured. In quantum physics, this process has come to be known as the *collapse of the wave function*: the measurement collapses the combined state into a state with a single eigenvalue. This mechanism has generated various philosophical interpretations, one of which—advocated by Wigner and von Neumann—requires the injection of consciousness in the act of measurement and its outcome.

In his book, *The Mathematical Foundations of Quantum Mechanics*, John von Neumann argues that the mathematics of quantum mechanics allows for the collapse of the wave function to be placed at any stage in the causal chain from the measurement device to the "subjective perception" of the human observer. In the 1960s, Eugene Wigner proposed that the consciousness of an observer is the demarcation line which precipitates the collapse of the wave function, independent of any realist interpretation. The non-physical mind is postulated to be the only true measurement apparatus. In short "through the creation of quantum mechanics, the concept of consciousness came to the fore again: it was not possible to formulate the laws of quantum mechanics in a fully consistent way without reference to the consciousness."[7]

Later on, Wigner took this outlandish claim back. However, to the young physicists of the 1960s, having been exposed to the Eastern theosophy brought to the West by the army of the swamis from India, Wigner's proclamation was a valuable endorsement of the scientific basis of the universal consciousness of Buddhism and Hinduism, to which the young physicists were attracted.[8]

Weyl's Consciousness

Hermann Weyl, the great German mathematician who joined the IAS in 1933 after the Nazis assumed power in Germany, was also an influential figure in theoretical physics. He was instrumental in introducing the physical ideas of quantum theory and relativity to mathematicians and informing physicists of the significance of the mathematical ideas of early twentieth century in physics. He described himself as an "unwelcome messenger" between the two communities—reflecting the mutual disregard of the two communities that was present in the early decades of the last century. Like many early contributors to quantum physics, Weyl fell under the spell of mystical doctrines. In his popular textbook on general relativity, Weyl writes, "the real world, and every one of its constituents with their accompanying characteristics, are, and can only be given as, intentional objects of acts of consciousness,"[9] and he opens his 1934 Yale lectures on 'Mind and Nature' claiming "the mathematical-physical mode of cognition … is decisively determined by the fact that this world does not exist in itself … [but] only as that met by an ego."[10]

Wheeler's Consciousness

John Archibald Wheeler, arguably one of the most influential theoretical physicists of the second half of the twentieth century, was smitten not only by the quantum-physics-mysticism connection, but also by the ambush of personal computers and the subsequent explosion of interest in information theory. After receiving his PhD from Johns Hopkins University at a time when American physics was beginning to blossom, Wheeler went to Europe to study under Niels Bohr. Together, they explained the mechanism behind nuclear fission.

Wheeler is credited with the revival of the general theory of relativity, which, in the race for nuclearization after WWII and the emergence of the new field of elementary particle physics, lay dormant well into the 1960s. He is also credited with the now ubiquitous terms like "black hole" and "wormhole." Although Wheeler did not win a Nobel Prize, he supervised 46 PhD students at Princeton University two of whom won the prize: Richard Feynman for his contribution to quantum electrodynamics, and Kip Thorne for his role in the design and construction of the LIGO detector and the observation of gravitational waves. All these achievements did not take away anything from Wheeler's modesty and his willingness to help the novice. He organized seminars for every entering class of graduate students in the Physics Department at Princeton University and personally supervised them to do research on a topic of their choice and helped them give presentations on their findings.[11]

There was another side to Wheeler, which often generated in him—perhaps because of his association with Bohr—an urge to speculate on matters outside science. Freeman Dyson, a long-time friend and a fellow speculator, describes John Wheeler as both prosaic and poetic. "In the fission paper, I met the prosaic Wheeler, a master craftsman using the tools of orthodox physical theory to calculate quantities that can be compared with experiment. The prosaic Wheeler has his feet on the ground. ... But from time to time, we see a different Wheeler, the poetic Wheeler, who asks outrageous questions and ... writes books with titles such as 'Beyond the black hole,' 'Beyond the end of time,' and 'Law without Law.'"[12]

The poetic Wheeler is the creator of an unsubstantiated conjecture with the koan-like epithet: "It from Bit." Wheeler treats the click of a counter when detecting a particle as a "yes," and its absence a "no." He then speculates that every physical quantity, every "it," derives its ultimate significance from bits, binary yes-or-no indications, and goes on to elaborate on this conjecture by speculating further:

> It from bit symbolizes the idea that every item of the physical world has at bottom – at a very deep bottom, in most instances – an immaterial source and explanation; that what we call reality arises in the last analysis from the posing of yes-no questions and the registering of equipment-evoked responses; in short, that all things physical are information-theoretic in origin."[13]

Two points in the quotation above pop up as pseudoscientific. The first one is the use of the word "deep." It sends a mystical signal so common in the writings of self-help gurus and mind-body mystics: a prayer can be effective only if the praying is done in a very "deep" state of mind; you can achieve all your goals by simply thinking "very deeply" about them; the ability of cancer survivors who have learned the mind-body connection springs from a level so "deep" that you cannot go any deeper. The second point is the use of the word "immaterial." Immaterial (or non-material) is the essence of pseudoscience. By New Age healer's own admission, objects like energy healing, Qi field, and reiki are non-material.

Contrary to Wheeler, information is a manifestation of matter. Wheeler's yes-no (or zero-one) binary does not exist without a material object registering them. Underneath any occurrence of a zero-one binary is a *material* switch that turns on (one) or off (zero). The switch could be a vacuum tube, as in the first generation of computers, a discrete transistor, as in the next generation of computers, or a microchip, as in modern electronic devices. Information without a material object carrying it is an impossibility not unlike the impossibility of energy—claimed by spiritualists to be non-material—without a material particle carrying it.

No mystical interpretation of reality is complete without consciousness, and Wheeler does not hesitate to pay due respect to it. Wheeler challenges Marie Curie's assertion, "Physics deals with things, not people," by disputing his own version of it:

> Using such and such equipment, making such and such a measurement, I get such and such a number. Who I am has nothing to do with this finding. Or does it? ... any claim to have 'measured' something falls flat until it can be checked out with one's fellows.[14]

If Schrödinger's consciousness resided in a single conscious person, Wheeler's exhibits itself in a community of conscious people: the hotness of the boiling water does not exist unless the reading of the thermometer is communicated with, and checked out by, "one's fellows". Generalize this to the existence of trees, clouds, Moon, and stars that came into being hundreds of millions of years before any of our "fellows," and you'll see the absurdity of the assertion.

Wheeler was also a powerful voice for the *Anthropic Principle*, which places human beings at the center of the universe.[15] It asks the question, "Why do physical constants have the value they have?" and proposes the answer: "If the constants were slightly different, the universe would not be able to create mankind." The answer is misguided and incomplete, because the physical constants are crucial not only for the emergence of life, but also for the formation of atoms, molecules, planets, stars, and galaxies in their present shapes and forms. The creation of mankind, on the other hand, requires much more than just the right values of the physical constants. It needs a planet that is solid enough, has sufficient amount of water, is at the right temperature, has appropriate chemical structure, and a lot more. The physical constants had the same values sixty million years ago when there was no sign of even primates.

Promoters of the Anthropic Principle go beyond the requirement of specific physical constants for the emergence of mankind. They assert that the constants have *been fine-tuned* so that *homo sapiens* could emerge. The fine-tuning must, therefore, have been done by a designer, a universal consciousness, a deity, or a God. In Wheeler's words:

> It is not only that man is adapted to the universe. The universe is adapted to man. Imagine a universe in which one or another of the fundamental dimensionless constants of physics is altered by a few percent one way or the other. Man could never come into being in such a universe. This is the central point of the Anthropic Principle. According to this principle, *a life-giving factor lies at the centre of the whole machinery and design of the world*.[16] [Emphasis added]

A universe "in which one or another of the fundamental dimensionless constants of physics is altered by a few percent one way or the other" will look completely different, not because it doesn't give life to mankind, but because it may not even have galaxies, stars, or planets as the present universe has. Therefore, associating the present values of the physical constants to the creation of mankind by a "life-giving factor" is a theological principle, plain and simple. One might as well replace "life-giving factor" with God, intelligent designer, universal consciousness, universal energy, or any other phrase invented by modern spiritualists.

The words of physicists of Wheeler's caliber become edicts to scholars whose expertise lies outside of sciences. An academician specializing in the "psychology of spirituality" refers to "It from Bit" as scientific evidence for the claim that consciousness—which he identifies as information—creates matter.[17] *It from Bit* is indeed an intriguing idea that can serve pseudoscientists and social scientists alike by providing an alleged quantum mechanical basis for their speculations. But John Wheeler's service to pseudoscientists goes beyond that.

Experimenter's Consciousness

In his *It from Bit* article, Wheeler introduces the term "observer-participant," alluding to the quantum physics founders' faulty notion that the outcome of a measurement depends on the observer that makes that measurement. He promotes a "participatory universe" in which observers of a scientific experiment participate in—and therefore affect—the result of that experiment. This has come to be known as the *experimenter effect* or *observer effect* in pseudoscientific literature.

Elsevier is a well-respected publisher in the scientific community. Among its journals are *Physics Letters* and *Nuclear Physics* in which high quality articles have been published over the years, some of which have earned their authors prestigious awards including the Nobel Prize. The same Elsevier started a "peer-reviewed" journal in 2005 whose executive, associate, and assistant editors all have pseudoscientific tendencies, and all submitted articles are written and reviewed by pseudoscientists. The journal is called *Explore: The Journal of Science and Healing*. The subtitle of the journal exposes its dubious identity: How can a journal publish healing *and* science—not medical science, health science, or hygiene science, but science, period—at the same time? A cursory look at the articles available online demonstrates that speculations about consciousness and cosmology, spiritual phenomena and quantum non-locality, and scientific investigation of reincarnation are legitimate candidates for publication.

Larry Dossey, the Executive Editor of *Explore*, is the author of books like *The Power of Premonitions: How Knowing the Future Can Shape our Lives* and *Prayer is Good Medicine*. Once in a while, he writes a column in *Explore* entitled "Exploration." One of these is devoted to "research" in the healing power of prayer. Dossey, an authority on faith healing, offers a list of twenty suggestions on the future research of prayer healing. He starts the list by criticizing the well-established practice of double-blind protocol to eliminate the placebo factor: "Experiments involving prayer should replicate, not subvert, how prayer is employed in the daily lives of ordinary people. Therefore, it is time to question whether the randomized double-blind protocol favored in conventional clinical research is adequate for healing experiments." Instead of randomized double-blind protocol, he encourages the inclusion of testimonials—a notoriously flawed procedure—as scientific evidence for the efficacy of prayer in healing.

But his most egregious statement is his third suggestion in which he uses John Wheeler's participatory interpretation of quantum physics:

In view of the evidence for experimenter effects, the preexisting beliefs of prayer experimenters should be ascertained and recorded as part of the study.[18]

Is Dossey implying that the results of studies that negate the efficacy of prayer in healing could be attributed to the incredulity of experimenters in prayer healing?

A group of mystics with backgrounds in biology, neuroscience, psychology, medicine, and psychiatry (with no representation from physics or chemistry) held a summit on post-materialist science, spirituality, and society, the outcome of which was a "manifesto" ... of the kind that political parties—such as the Communist party of the mid-nineteenth century under the leadership of Marx and Engels—write to proclaim their agenda and suggest course of future action. Larry Dossey signed on to the article as an author and published it in *Explore* as a Guest Editorial. The "Manifesto for a Post-Materialist Science" identifies the weakness of science as being based on materialism and reductionism, to which it attributes the stagnation in the development of the scientific study of mind and spirituality. "Manifesto" notes the discovery of quantum mechanics (QM) in the 1920s and 1930s, and proclaims that

> QM explicitly introduced the mind into its basic conceptual structure since it was found that particles being observed and the observer – the physicist and the method used for observation – are linked. According to one interpretation of QM, this phenomenon implies that the consciousness of the observer is vital to the existence of the physical events being observed and that mental events can affect the physical world.[19]

The voices of Bohr, Heisenberg, Wigner, and Wheeler are disturbingly audible in this passage. "Manifesto" goes on to say that "The results of recent experiments [on psi phenomena, telekinesis, extrasensory perception (ESP), near-death experience (NDE), out-of-body experience] support this interpretation. These results suggest that the physical world is no longer the primary or sole component of reality and that it cannot be fully understood without making reference to the mind."

"Manifesto" suggests that paranormal phenomena appear anomalous only when seen through the lens of scientific materialism. The fact that psi phenomena cannot be seen, felt, heard or measured by any instruments is only an anomaly. If we accept the post-materialistic argument that—lamentably in accordance with the mystical writings of some great physicists—mind can influence the outcome of a physical experience (that imagination substitutes actual physical measurement), no anomaly will exist and psychic phenomena, and pseudoscience in general, will advance to the level of science!

Admission of Guilt

While New Age gurus call the mystical views of notable physicists at the top of their voices and lecture their followers on the scientific basis of their mystical hogwash, they never even whisper anything about the same physicists' subsequent rejection of those very views.

Dyson Takes It Back: Among countless wild ideas that Freeman Dyson has speculated, his principle of maximum diversity, a derivative of the Anthropic Principle, is noteworthy. It says that "the laws of nature and the initial conditions are such as to make the universe as interesting as possible. As a result, life is possible but not too easy. ... Examples of things which make life difficult are all around us: comet impact, ice ages, weapons, plagues, nuclear fission, computers, sex, sin, and death. ... Maximum diversity often leads to maximum stress."[20] When pressed for comments on this principle, Dyson said that he didn't intend anyone to take that too seriously and that

> I never think of this as a deep philosophical belief. It's simply, to me, just a poetic fancy.[21]

Wigner Takes It Back: The Wigner of the 1960s, who declared that it was not possible to formulate the laws of quantum mechanics without reference to consciousness, started to change his position in the 1970s. In a paper that he published in 1984, he wrote,

> This writer's earlier belief that the role of the physical apparatus can always be described by quantum mechanics ... implied that the collapse of the wave function takes place only when the observation is made by a living being [a *conscious* person] – a being clearly outside the scope of our quantum mechanics. The argument which convinced me that quantum mechanics' validity has narrower limitations, that it is not applicable to the description of the detailed behavior of macroscopic bodies [such as a conscious person], is due to D. Zeh."[22]

6 Quantum Consciousness Crosses the Atlantic

Weyl Takes It Back: On the fact that he shared the philosophical premise that postulated consciousness as the foundation of physical reality, Weyl said

> I was too prone to mix up mathematics with physical and philosophical speculation.[23]

In the hope of exposing the ruse of New Age gurus who want to mix the message and the messenger, I close this chapter—as I did in Chap. 3—with repeating, for every physicist who believes in mysticism, what Einstein said about Eddington's and Jeans' "literary formulation" versus their "scientific conviction":

– Dyson's literary formulation that "the universe shows evidence of the operations of mind on the level of elementary physical processes in quantum mechanics" must not be taken as expressive of his scientific conviction.
– Wigner's literary formulation that "it [is] not possible to formulate the laws of quantum mechanics in a fully consistent way without reference to the consciousness" must not be taken as expressive of his scientific conviction.
– Weyl's literary formulation that "the real world, and every one of its constituents can only be given as intentional objects of acts of consciousness" must not be taken as expressive of his scientific conviction.
– Wheeler's literary formulation believing that "The universe is adapted to man and a life-giving factor lies at the center of the whole machinery and design of the world" must not be taken as expressive of his scientific conviction.

Pseudoscience has acquired considerable might since the days when it was just a curiosity among a small group of science enthusiasts with an unbridled imagination. Thanks to the philosophical and mystical speculation of great physicists, pseudoscience has been emboldened to the point that it now dares to undermine the methodology and essence of science, as exhibited by the "Manifesto." The emboldenment would have been unlikely were it not for the social revolution that gripped the West as a result of an unpopular war in the East.

7

Eastern Plague of the Sixties

The social upheaval of the 1960s in America was a mixed blessing. The success of the Civil Rights Movement, the antiwar demonstrations that eventually brought the Vietnam War to a halt, and the Feminist campaign for gender equality, were three good things that came out of the chaos of that period. But there were also some ugly outcomes.

As the chants of Hare Krishna began to fill the air of the mid-1960s, and the pubertal youth realized that in lieu of the Western religions' teaching of abstinence, some Eastern theosophies taught the worship of gigantic statues of phalluses, Western floodgates opened up to Eastern mystics and the "*guru*" quickly became a household word in the U.S. Some gurus came and went quickly, often amidst scandalous debauchery seasoned with illicit drugs. Others settled into the American landscape, where their influence is still felt today. The antiwar sentiment of the youth became an agent of recruitment for the gurus who portrayed themselves as messengers of peace.[1]

The tranquil smiling face of a monk in namaskar gesture was the palpable antithesis of the Vietnam War. This contrast was especially manifest on campuses where the covert association of universities with the military enraged students and faculty. Sit-ins, boycotts, and occupation of administration buildings dominated the academic life on countless campuses. The misconceived association of hard sciences—rather than the *technology* derived from them[2]—with the War initiated an anti-science wave encapsulated in the phrase "military-industrial complex." In their search for a solution to this anti-science sentiment, into which the youth was being sucked, some physicists contemplated on neutralizing the militancy associated with their field by putting the seemingly peaceful face of Eastern theosophy on it.

© The Author(s), under exclusive license to Springer Nature Switzerland AG 2024
S. Hassani, *Quanta in Distress*, https://doi.org/10.1007/978-3-031-65259-2_7

Institute for Advanced Mysticism

The institutions of higher learning of the West Coast, most notably the University of California at Berkeley, and Stanford University, were not only the hotbed of resistance, but also the launching pad of the New Age Movement. Michael Murphy, during a seated meditation by Lake Lagunita at Stanford, experienced what he described as a "hinge moment", at which point he decided to leave the premed program in which he had enrolled, get a degree in psychology and pursue meditation in India. Dick Price, a Stanford psychology graduate, during a manic psychosis episode for which he was hospitalized, came to fantasize that the episode, which he referred to as "the state", was actually the healing process from what went on with him prior to it. Murphy, upon returning from India, came across Price and the two decided to found the Esalen Institute in Big Sur, California to support alternative methods for exploring human consciousness. On its website, Esalen claims that it "has proven the possibilities of reconciling intellectual and experiential; mind and body; science and mysticism; immanence and transcendence; East and West; meditation and action; youthful idealism and time-tested wisdom."[3]

At Berkeley, the New Age movement of the Sixties took a "scientific" turn. Two physics graduate students, Elizabeth Rauscher—who, the last I checked, was doing research in psychic healing, faith healing and other paranormal claims[4]—and George Weissmann, dissatisfied with lack of emphasis on philosophy in the physics courses taught at Berkeley, founded the Fundamental Fysiks Group[5] in San Francisco to explore the philosophical implications of quantum theory. Leading members included

- Fritjof Capra, PhD, of whom more will be said shortly;
- Nick Herbert, PhD, who constructed a "Metaphase Typewriter", a supposedly quantum device whose purpose was to communicate with disembodied spirits, but despite many tests, including an attempt to contact the spirit of Harry Houdini on the hundredth anniversary of his birth, no success was achieved with the device;[6]
- Jack Sarfatti, PhD, whose website carries updates on recent UFO sightings, spotting Bigfoot in Argentina and UK and how Bigfoot saved a family from tornado wreckage;[7]
- Henry Stapp, PhD, who argues that quantum wave functions collapse when conscious minds select one among the alternative quantum possibilities and has coauthored an article with Deepak Chopra on the Huffpost arguing that consciousness is fundamental to the universe;[8] and

— Fred Alan Wolf, PhD, who writes popular books on the connection between quantum physics and spirituality, the Yoga of time travel, alchemy of science and spirit, the unity of psyche and physics, shamanism and physics, and the physics of mind-body and health.[9]

The intellectual activity of the frequenters of the Esalen Institute and the Fundamental Fysiks Group was bound to reach the larger audience outside, eagerly awaiting a justification for their abandonment of traditional beliefs of the West in favor of the new thought brought by the peace-loving gurus of the East. And what better reason to join the ashrams of gurus than if their new belief system was based on the exact science of physics. Two books pioneered the mixing of Eastern theosophy and quantum physics. They were both influential in creating the faulty popular mindset that there was indeed a connection between Eastern mysticism and quantum physics and laid the foundation on which the trivialization and degradation of quantum physics that we see today could be erected. One of these books was written by a physicist who regularly attended the Fundamental Fysiks Group, the other by a lay person who was introduced to a mystical version of physics by the people at the Esalen Institute.

Dancing Shiva and Wu Li Masters

As he was "sitting by the ocean one late summer afternoon [in 1969, he] saw cascades of energy coming down from outer space, in which *particles were created and destroyed* in rhythmic pulses; I 'saw' the atoms of the elements and those of my body participating in [a] cosmic dance of energy; I felt its rhythm and I 'heard' its sound, and at that moment I *knew* that this was the Dance of Shiva, the Lord of Dancers worshiped by the Hindus."[10] [Emphasis added] With these words, Fritjof Capra sets out to establish his alleged parallelism between modern physics and Eastern mysticism in his influential book, *The Tao of Physics*.

This subjective, personal, unverifiable experience is not unlike the epiphany alleged by spiritual leaders, who after leading a sinful life and committing illicit acts, claim to have had "a spiritual awakening" during a court trial or while serving their terms in prison. And the public falls for it time after time. Capra's confession is much more effective because it carries the convincing weight of modern physics.

Capra has a background in high energy physics whose mathematics contains "creation and destruction" operators. He puts these words next to "cosmic dance of energy," a notoriously Eastern concept, to induce his readers to accept the alleged parallelism between physics and Eastern thought. This ruse of "proximity implies parallelism" is very common in mystics' literature, and we will encounter it again and again later.

Reality of the Unreal

Physics relies on quantifiable observations and experiments using measuring devices—which in their most primitive forms were our senses—and drawing rational conclusions (theories) from them. Eastern thought speaks of a direct experience of reality which transcends not only intellectual thinking but also sensory perception. Buddhists call the knowledge that comes from such an experience "absolute knowledge."

> The Eastern mystics repeatedly insist on the fact that [the absolute knowledge of] the ultimate reality can never be an object of reasoning or of demonstrable knowledge. It can never be adequately described by words, because it lies beyond the realms of the senses and of the intellect from which our words and concepts are derived.[11]

The passage would cry out its emptiness if you replaced "the ultimate reality" with "ghosts", "Santa Claus", "leprechauns", "angels", "devil", "God"—or any other entity that "lies beyond the realms of the senses and of the intellect"—with minor change in the structure of the narrative and note that the new sentence makes as much (or as little) sense as Capra's. I'll do "Santa Claus" as an example; you can try any other entity that is beyond our senses:

> The believers in Santa Claus repeatedly insist on the fact that Santa Claus can never be an object of reasoning or of demonstrable knowledge. He can never be adequately described by words as to how he climbs millions of chimneys in one night, because he lies beyond the realms of the senses and of the intellect from which our words and concepts are derived.

What is a reality that cannot be seen, felt, heard, or touched? It is an unverifiable entity that can take as many shapes and forms as there are people trying to observe it. And since it cannot be described by words, no two human beings can check whether their realities are the same. This reality is the first half of the title of Capra's book.

7 Eastern Plague of the Sixties

The reality of the second half of the title is the antipode of Tao. Physics describes a veritable reality that is global, even universal, but not eternal. It is a reality that has rid itself of theosophies, traditions, and folklore. It is a reality, which—once they go through the necessary training—every individual of our race, regardless of their origin and background, can grasp and concretely verify and communicate to their peers. In its current form, this universal reality consists of a handful of fundamental particles, which make up the entire universe, which itself was created in a big bang 13.78 billion years ago. This reality informs us that a few fundamental particles merge together to form atoms; that the two lightest atoms, hydrogen and helium, make up over 99% of the visible universe. It is a reality created by the irrefutable force of observation, and that force tells us that the universe is expanding faster and faster.

That reality combines atoms to form molecules, a form of matter most abundant on Earth. Some of these molecules are simple and consist of a few atoms. Some are very complex and composed of thousands of less complex molecules, each of the latter carrying hundreds of atoms. Some of these very complex molecules were formed a few billion years ago and had a property which we now call *life*.

The reality of physics—and of science, in general—is a dynamic reality which evolves as the scientific wisdom of our race evolves. It is discovered through the laborious undertakings of generations of scientists who *communicate* their discoveries not only to their peers, but more importantly, to the next generation of scientists, who build on the knowledge of all the previous generations. And this communication is crucial for the development of science and the recognition of reality.

A child *knows* that there are monsters under his bed; a literal believer in Bible *knows* that God created the Earth on September 17, 3928 B.C.; a Christian fundamentalist *knows* that blowing up an abortion clinic will send his soul to heaven after his death; a Trump follower *knows* that the 2020 election was rigged and that the devil worshiping, baby-blood sucking Democrats stole the election and that it is their duty to go to Washington to occupy the Capitol; Capra wants to make you *know* that reality is untouchable, unseeable, unknowable, and therefore, unreal. … Once you step outside the domain of science, fantasy and reality become indistinguishable.

Much has been said about the wisdom of Eastern philosophy. If there is any truth to that, one outstanding example of that wisdom is its prohibition of any form of communication. By not articulating the absolute truth and instructing the posterity not to talk about their experience, the historical masters of Eastern mysticism made their philosophy unfalsifiable. They were clever enough not to make the same mistake made by their counterparts farther to the west, where a group of sages made the biggest blunder of history by claiming that God created the universe in six days; and today millions of followers are scratching their heads trying to figure out what exactly "six days" means.

Conscious Photon?

Gary Zukav appeared 34 times on *The Oprah Winfrey Show* and one of his books was the #1 New York Times bestseller 31 times and remained on the list of bestsellers for three years. For a gullible, scientifically illiterate public, these credentials make Zukav, like his mentor Oprah, an icon to follow.

Zukav had no connection with the scientific community until one of his friends invited him to a conference at the Esalen Institute. "To my great surprise, I discovered that (1), I understood everything that they said, and (2), their discussion sounded very much like a theological discussion. ... [Physics] was a rich, profound venture which had become inseparable from philosophy."[12] Contrast this with the typical reaction of a physicist attending a professional conference: "I didn't understand a word of what they were talking about." This reaction speaks of the effort of physicists at understanding new concepts, which—in contrast to political, philosophical, or theological ideas—does not come while listening to a speaker at a conference but through laborious examination of the proposed ideas afterwards.

Who are the physicists that attend a conference in which a layman like Zukav could understand everything they say? In the acknowledgement of *The Dancing Wu Li Masters*, Zukav expresses his indebtedness and gratitude to several people. Most notable among them are the regular attendees of the Fundamental Fysiks Group's meetings in San Francisco.

Zukav's book is filled with quotations by well-known mystical physicists. After a quote by John Wheeler, in which the observer becomes entangled in the observed, and a quote by Carl Jung (the Swiss psychoanalyst famous for his mystical viewpoint)[13] in which the psyche becomes part of the physical world, Zukav concludes "If these men are correct, then physics is the study of the structure of consciousness."[14]

Are there any signs of "consciousness" in what physics studies? To find out, let's go back to the double-slit experiment. We saw that when each slit is open by itself, a blob appears on the photographic plate. However, when we open both slits and wait until a sufficient number of photons have passed through them, instead of two expected blobs we see a pattern of bright and dark fringes. This strange behavior, as we saw in Chap. 5, can be explained by the probabilistic nature of quantum phenomena. But Gary Zukav has his own explanation:

> *The question is, How did the photon in the first experiment know that the second slit was not open?* Think about it. If both slits are open, there are *always* alternating bands of illuminated and dark areas. This means that there are always areas where the photons never go ....
> When we fired our photon and it went through the first slit, how did it "know" that it could go to an area that must be dark if the other slit were open? In other words, how did the photon know that the other slit was closed?
> There is no definite answer to this question. Some physicists, like E. H. Walker, speculate that photons may be *conscious!*[15]

The assumption of the photon being conscious is evident from the beginning. Using the word "know" for the photon, Zukav already imparts the capability of making choices to it, rendering a photon intelligent and conscious. Is there really "no definite answer to this question" as Zukav wants us to believe? Of course there is! Chapter 5 showed us that there is a perfectly reasonable answer to the question if you are open-minded enough to accept the fact that quantum physics is probabilistic, and you cannot "explain" probability without falling into the trap of consciousness.

As for Zukav's "physicist", E. H. Walker, a search on the internet informs us that Walker received his Ph.D. in 1964 from the University of Maryland, but there is no indication that he worked at any institution of higher learning. He wrote a book called *The Physics of Consciousness* and received the Outstanding Contribution Award from the Parapsychological Association in 2001 for his quantum theory of consciousness. Parapsychology is a pseudoscientific discipline that studies psychic abilities, near-death experiences, out-of-the-body experiences, crisis apparitions, retro-cognitions, reincarnation memories, regression memories, prophecy, astrology, ghosts and life after death, all grouped under the umbrella of *psi phenomena*.

Attempts at explaining *any* probabilistic outcomes sends the explainer down the drain. It is worth repeating the coin experiment of page 40: If you toss 10 coins, the probability of getting 6 heads (60% of the total) is about 0.205. If you toss 10,000 coins, the probability of getting 60% heads is 0.000...00029 (replace the dots with 83 more zeros). Now regard the 10,000 coins as 1000 groups of 10 coins. Each group, in isolation, has a 20.5% chance of getting 60% heads, but the presence of other groups renders getting 60% heads practically impossible. How does one "explain" this? Here is my shot at what Zukav might offer:

The question is, How does each group of ten coins know that it is part of 10,000 coins and therefore it should avoid showing too many heads, as it would if it were an isolated group of ten coins? Think about it. Each group of ten coins shows 60% heads 20.5% of the time. But once it becomes part of the other groups, practically no head shows up. Could it be that the members of each group of ten tell each other "Remember, we are now only one of 1000 teams. So, let's not show our heads?" Could it be that the coins are conscious just as photons are?

It is easier to mystify what we can't see than what we have daily experience with. We know that a coin cannot "know," but we are not sure if saying that a photon "knows" is such a crazy idea. In fact, it seems to be very appealing, as we are so fond of seeing life in inanimate objects in a Disney movie: as long as the world of photons is invisible, why not make it as fanciful as possible. However, the truth is that, since both the crazy and the not-so-crazy ideas have identical probabilistic explanation, it is just as absurd to say that a photon is conscious as to say that a coin is conscious.

Classical Physics and Its Umbilical Cords

The underlying characteristic of modern physics is its inaccessibility to our senses: relativistic effects for objects that move at—even the largest—speeds we ordinarily experience are so small that only the most technologically advanced devices can measure them; the domain in which quantum physics applies is the world of subatomic particles, which are orders of magnitude removed from our experience with even the smallest ordinary objects. Modern gurus exploit this inaccessibility to align their mystical beliefs with modern physics. Their argument goes something like this: you can't experience modern physics; you also can't experience Eastern mysticism because it cannot be heard, seen, smelled, tasted, and no words can describe it. Therefore, there is a parallel between modern physics and Eastern mysticism.

Classical physics, on the other hand, has no commonality with Eastern mysticism because it can be sensorially experienced. And if that is the case, then there should be no commonality between classical physics and the partner of Zen, Buddhism, Taoism, and Hinduism: modern physics. It is therefore imperative for Eastern mystics to drive a wedge between modern and classical physics. They use "mechanistic" to describe classical physics and have been successful in demonizing the word for the public and the untrained academicians. Here is how Zukav assesses the contrast between Wu Li and

Newtonian physics on the one hand and its similarity to twentieth century physics on the other:

> Wu Li, the Chinese word for physics, means 'patterns of organic energy' …. This is remarkable since it reflects a world view which the founders of western science (Galileo and Newton) simply did not comprehend, but toward which virtually every physical theory of import in the twentieth century is pointing![16]

Even though the purpose of the quote is to snatch modern physics away from classical physics, Zukav's attempt is muddled. Is Wu Li the Chinese word for physics or *modern* physics? Do Chinese consider classical physics as "physics" or something else? If Wu Li means 'patterns of organic energy'—the notorious telltale sign of mysticism—and it is the word for physics (in general), then should we not assume that the physics of Galileo and Newton also point to a world view that is reflected in 'patterns of organic energy'? But we'll be charitable and forgive Zukav for his confusion.

Is modern physics really the antithesis of classical physics? The development of the two pillars of modern physics, quantum mechanics and relativity theory, shows that although modern and classical physics differ, their difference is more like the dissimilarity between a mother and her twin daughters: the latter are umbilically attached to the former and cannot come into being without her.

Starting in 1820, signs of the unity of electricity and magnetism—hitherto assumed to be different forces—began to emerge. By 1865, four mathematical equations summarized all discoveries—*made entirely in classical physics settings*—in electromagnetism, as the field came to be known by then. As James Clerk Maxwell, a *classical physicist*, examined those equations, he realized that they were mathematically inconsistent, and, by changing the fourth equation, he removed the inconsistency, and earned the honor of attaching his name to the new four equations. Further mathematical manipulation revealed to Maxwell that there must exist electromagnetic waves (EMWs) *always* traveling at the speed of light, denoted by c. From this conclusion, two other *classical physicists* erected the two pillars of modern physics.

Max Planck, a *classical physicist* in December 1900, while studying how heat produces EMWs, discovered the formula for a curve called the *black body radiation curve*. The BBR curve was a fit to the observation points of the plot of intensity of the EMWs as a function of their wavelength (color). The observations and the measuring devices involved were all done using *classical physics* by *classical physicists*. To explain his curve, Planck invented the idea of quantum, thereby laying the foundation of one pillar of modern physics.

Albert Einstein, a *classical physicist* in June 1905, while studying EMWs in moving bodies and the fact that the speed of an electromagnetic wave does not depend on the motion of its source,[17] concluded that time is affected by motion and that observers traveling relative to one another experience different times. Thus, Einstein laid the foundation of the other pillar of modern physics.

That the two pillars of modern physics are simply extensions of classical physics can be established by the following additional observation. Classical physics is not suited for studying objects that are of sub-atomic size—where quantum physics ought to be used—and/or move close to the speed of light—where relativity becomes relevant. On the other hand, when a number that, for small values, describes quantum phenomena gets large, it turns into a description of classical physics. Similarly, for slow objects, the formulas of relativity yield the corresponding classical physics formulas. The last two sentences encapsulate the so-called *correspondence principle*.

It pays to repeat the following quote, in a mantra-like fashion, to New Agers: "*Quantum physics and relativity were born out of the womb of classical physics. The former are extensions of the latter. Without classical physics there would be no quantum physics or relativity.*"

Perversion of $E = mc^2$

In the years after the discovery of relativity in 1905, Einstein and other physicists and mathematicians developed relativity to a point where it became the cornerstone of all of *relativistic physics*. As a result of this development, classical concepts such as energy and momentum underwent substantial improvement to make them consistent with relativity. One of the byproducts of this improvement is the famous equation, $E = mc^2$, which has been gravely abused by Eastern mystics, who identify the non-material soul and spirit with energy, the left-hand side of the equation. This identification is the result of assuming that energy is non-material—a false assumption made even by some professional physicists. The equation then becomes a scientific proof of the equivalence between soul and matter:

> In the East ... there never has been much philosophical or religious ... confusion about matter and energy. The world of matter is a relative world and ... we do not see it as it really is. The way that it really is cannot be communicated verbally, but in the attempt to talk around it, eastern literature speaks repeatedly of dancing energy and transient, impermanent forms.[18]

Dance can be attributed to any kind of movement, and the East is notorious for the abundance of these movements. You can label the motion of subatomic particles or their creation and annihilation as "dance," but such labeling, no matter how poetic you make them sound, does not prove any similarity between Eastern mysticism and the world of subatomic particles.

While the Eastern matter "cannot be communicated verbally," mass m can not only be communicated verbally, but observed and measured with mathematical precision. And $E = mc^2$ is not some mystical "equivalence" of mass and "dancing energy," but a measurable transformation of one into the other.

What exactly is energy? Consider kinetic energy (KE), the energy associated with the motion of an object. It is given as a formula in terms of the velocity of that object. Asking whether or not KE is material is tantamount to asking whether or not velocity is material. You can see the absurdity in even phrasing the question. Velocity is an observable and measurable *property* of matter in motion. A red apple, a black sheep, or a white daisy is material. Does it make sense to say that redness, blackness, or whiteness is non-material? This confusion is a common pitfall in which even trained physicists can fall, and a dangerously effective tool that modern gurus use to promote their nonsense.

An example of the transformation of mass into energy is nuclear fission. A slow neutron hits the nucleus of uranium-235 and splits it into two lighter nuclei (called *daughter nuclei*), photons, and three neutrons. If you add the masses of the initial neutron and uranium-235, you'll find that it is larger than the total mass of the daughter nuclei and three final neutrons (photons are massless). The "missing mass" is converted into the energy of the photons and the kinetic energy of the two daughter nuclei and three neutrons: adding these energies will give the same result as the product of the missing mass times c^2. Therefore, energy is not some kind of a mystical entity on its own. It is a measurable and quantifiable property of the material daughter nuclei, the three material neutrons, and the material (but massless) photons.

Massless particles are favorite objects of mystics stolen from modern physics and mutilated beyond recognition. "You can see [energy transforming into mass] happening in elementary particle processes. A photon is transformed into two material particles: an electron and an antielectron. Material is produced from pure energy, from a photon."[19] This quote is an epitome of misunderstanding modern physics, even though it is asserted by an astrophysicist. A (single) photon can never create an electron and its antiparticle (positron) because photon is not some kind of formless, metaphysical, pure energy. It is a particle that has momentum. And conservation of momentum does not allow the process to take place. You need at least two photons to create an electron-positron pair.[20]

The E of $E = mc^2$ is *always* the energy of two or more real particles (with or without mass) that can either produce the m of the equation by binding themselves together, or be produced by the m as it decays into two or more real particles. There is no instance in nature in which mass transforms into energy (or vice versa) without some real observable particles carrying that energy. There is no connection between soul-matter equivalence of Eastern mysticism and energy-mass transformation of modern physics. Period!

That photon has no mass gives New Age gurus yet another opportunity to connect modern physics and Eastern mysticism:

> A "massless" particle is the name [physicists] give to an element in a mathematical structure. ... it is impossible [to describe that element] because the definition of an object (like a "particle") is something that has mass.
> Zen Buddhists have developed a technique called koan ... A koan is a puzzle which cannot be answered in ordinary ways because it is paradoxical. "What is the sound of one hand clapping?" is a Zen koan. Zen students are told to think unceasingly about a particular koan until they know the answer. There is no single correct answer to a koan. It depends on the psychological state of the student. ...
> Physics is replete with koans, i.e., "picture a massless particle."[21]

To compare a massless particle like photon, whose physical properties make it as unique as, and at the same time similar to, any other particle, with the sound of one hand clapping, which—despite its thought-provoking folly—"depends on the psychological state of the student," is a gross disfiguration of physics. And to say that the definition of a particle "is something that has mass" is either an indication of ignorance of what a particle is, or a ruse to coerce the reader into accepting the parity of physics and mysticism stated in the last paragraph. In 1939, Wigner showed, with the exactness of mathematics, that particles *can have zero mass* and he shared the 1963 Nobel Prize in Physics for that.[22]

Eastern mystics and pop-spiritualists may take delight in attributing paradoxes to modern physics to align it with their belief system. But if there are paradoxes in modern physics, it is only because we try to understand a physical phenomenon on the basis of our limited, incomplete, and mostly wrong intuition. The following is an example of a paradox in relativity, which epitomizes all paradoxes in physics.

When relativity discovered the notion of length contraction of moving objects, there seemed to be a paradox, which came to be known as "the pole and barn paradox." A runner moving at almost light speed and carrying a pole enters a barn through the front door. A little later, an observer standing in the barn notices that the pole fits snugly between the front and back doors and concludes that the pole has the same length as the barn. The runner, on the other hand, with respect to whom the pole is stationary, sees the pole longer and the length of the barn shorter—because the barn is moving relative to the runner. He concludes that the pole is longer than the barn! Who is right? It turned out that the paradox was the result of our intuitive notion of absolute time. The barn observer sees the two ends of the pole coincide with the two doors of the barn simultaneously. The runner does not, because simultaneity is relative! He sees—as a simple relativistic argument can show—the coincidence of the leading end of the rod with the back door *before* the coincidence of the trailing end of the rod with the front door. So, as the front end of the rod is exiting the back door, the back end is outside the barn. The runner concludes that the pole is longer. Both observers are right! Because they measure different times. Paradox resolved! Relativity is right, and our intuition is wrong. There is absolutely no room for koans in physics. Every student of physics who is told to "think unceasingly about this paradox" will answer it—after he/she masters relativity—and the answer does not depend on "the psychological state of the student." It is a single unique answer obtained by all students from Albania to Zimbabwe.

As for the "paradox" of the one-handed clap, it is no paradox after all! Put a highly sensitive sound detector next to a single moving hand and you'll see the dial go off the chart as the air molecules collide with the palm of the hand and produce sound. What makes it a paradox is the limited sensitivity of our ears, Zen Buddhists' lack of understanding of the physical nature of sound waves, relying on the "wisdom" of thousand-year-old scriptures, and disinterest of New Age gurus in referring to elementary physics books—despite their universal availability—to grasp what a sound wave is.*

Infesting Modern Physics

The investigation of combustion coincided with the start of modern chemistry in the eighteenth century. Because a flame rises away from a burning sub-

---

*If I sound pedantic here it is because of the insistence of New Age gurus to attach Eastern thought—which by itself could be poetically appealing—to modern physics. "What is the sound of one hand clapping?", by itself, is indeed a thought-provoking poetic question. But when it is put next to "picture a massless particle", to parallel Eastern thought with modern physics, it demands pedantry to annul the parallelism.

stance, it was thought that the burning process removed a substance, named *Phlogiston*, from the burnt material. The phlogiston theory of burning was so predominant in the early eighteenth century that when it was found that some objects actually gained weight after combustion, the advocates of the theory gave phlogiston a negative weight. The whole idea of phlogiston was eventually abandoned in light of its bizarre properties and strong experimental evidence against it.

Heat was another concept that started on the wrong track. It was assumed to consist of a fluid called *calorie*, which flew in and out of substances making them hot or cold. At some point, two kinds of calorie were in existence: positive (causing a rise in temperature) and negative (causing a drop in temperature). The caloric theory of heat eventually fell flat on its face when it was irrevocably shown that heat was a form of energy. Today, the phlogiston theory of combustion and the caloric theory of heat is of interest only to the historians and philosophers of science. A phlogiston/calorie-like theory popped up in the early 1960s.

The mushrooming of particle accelerators in the late 1950s and early 1960s flooded the physics community with hitherto unseen particles called *hadrons*. To make sense of the proliferation of hadrons, some physicists worked on a theory that was based on the assumption that hadrons were composed of more elementary particles, called *quarks*,[23] with some very strange properties—such as having electric charges that were a fraction of the charges of all other known particles.

Although the early quark model explained a variety of the observed properties of hadrons, the fact that quarks were never seen in isolation made it hard for some physicists to take the model seriously. If hadrons are made up of quarks, then by providing a sufficiently hard blow to a given hadron like proton, we should be able to knock one or more quarks out of it just as we can knock an electron out of an atom. But regardless of the strength of the blow, no isolated quark was ever observed. In all experiments, irrespective of the amount of the energy provided to the colliding hadrons, the final products were more hadrons.* Physicists with a mystical twist found hadrons begetting hadrons intriguing.

A seemingly attractive idea that caught the attention of physicists in the early 1960s was the *bootstrap hypothesis*. Since no constituent particles show up in the violent collisions of hadrons, argued Geoffrey Chew the originator of

---

*Chapter 10 discusses how the *Standard Model* of fundamental forces and particles explains this strange behavior.

the bootstrap hypothesis, hadrons themselves—being the byproducts of the interaction of hadrons—must be fundamental: hadrons are both fundamental and composite. This is strikingly similar to a Hindu parable:

> In the heaven of Indra, there is said to be a network of pearls, so arranged that if you look at one you see all the others reflected in it. In the same way, each object in the world is not merely itself but involves every other object and in fact is everything else.[24]

Although the bootstrap hypothesis survived into the early 1970s, it eventually lost its appeal due to its complicated assumptions, its inability to explain (in numerical terms) the outcome of hadron collision experiments, and its inability to predict any new physics—a property of any good theory.

The similarity between bootstrap theory and Eastern theosophy was, for modern gurus, too good to let go. Almost a quarter of *The Tao of Physics* is devoted to bootstrap and the related *S-matrix* theory. For Capra, hadrons begetting more hadrons becomes the interconnectedness of all hadrons, and—by a little mystical stretch—of all things:

> In the Eastern view then, as in the view of modern physics, everything in the universe is connected to everything else and no part of it is fundamental. The properties of any part are determined, not by some fundamental law, but by the properties of all the other parts.[25]

In this quote, Capra is not referring to the *Standard Model* which has explained all observable hadron interactions, has predicted many experimentally observed phenomena, and has brought tens of Nobel Prizes for its contributors. By "modern physics" Capra means the defunct bootstrap hypothesis.

Eastern theosophy abhors reductionism. The quark model wants to tear hadrons apart and reduce them to more fundamental entities. This is an antithesis of the holistic approach whereby the whole and its parts are united. So, when bootstrap theory said that hadrons simultaneously make up and are made up of hadrons, it was music to modern gurus' ears.

Chew never grew out of the idea of the bootstrap and got stuck in the mysticism that ensued from it. He became the cult leader of those New Agers who were looking for a holy grail in physics that would reinforce their mystical beliefs and found it in the bootstrap hypothesis. Physics is no longer Chew's profession. He believes that the future human intellectual endeavor will not be scientific at all: "Our current struggle [with certain aspects of advanced physics] may thus be only a foretaste of a completely new human intellectual endeavor, one that will not only lie outside physics but will not even be describable as 'scientific.' "[26]

On the other hand, those talented physicists who started with bootstrap, but abandoned it for the 'basic-building-block' theories, accomplished a lot. Here is what David Gross, one of Chew's former students, has to say, on the Nobel Prize web site, about his disillusionment with bootstrap and Chew:

> In 1964, I started to do research under the supervision of Geoffrey Chew, the charismatic leader of the S-Matrix "bootstrap" approach to the strong interactions. I found this revolutionary new theory very exciting at first, but gradually became disillusioned. I rapidly finished a thesis and spent most of my last year at Berkeley in thoughts of new directions.[27]

The "new directions" led Gross eventually to the discovery of quantum chromodynamics (QCD)—a component of the Standard Model.

Canine Mysticism and Field Theoretic Sanskrit

Space limitation does not allow a more detailed refutation of the alleged parallelism between physics and Buddhism, Hinduism, or Taoism. Nevertheless, a couple of "parallels" from Capra's book are irresistibly amusing. The first one has to do with the popular characterization of science as "repeatable:"

> Anybody who wants to repeat an experiment in modern subatomic physics has to undergo many years of training. Only then will he or she be able to ask nature a specific question through the experiment and to understand the answer. Similarly, a deep mystical experience requires, generally, many years of training under an experienced master, and, as in the scientific training, the dedicated time does not alone guarantee success. If the student is successful, however, he or she will be able to 'repeat the experiment.' The repeatability of the experience is, in fact, essential to every mystical training and is the very aim of the mystics' spiritual instruction."[28]

Some readers may have noticed the shallowness of the argument in the quotation above. This kind of argument is a very common—in fact, it is the majority of—syllogisms used by the mystics to establish a parallelism between Eastern thought and modern physics. They put a mystical statement next to a similar-sounding statement about science—or a quotation by a mystic scientist—and argue that the similarity of those statements implies the parallel between the contents.

7 Eastern Plague of the Sixties

For those readers who may have missed the emptiness of the reasoning above, the following argument for parallelism may shed some light on the fallacy of the syllogism:

> Anybody who wants to repeat an experiment in modern subatomic physics has to undergo many years of training. Only then will he or she be able to ask nature a specific question through the experiment and to understand the answer. Similarly, a sophisticated canine trick requires, generally, many years [okay, months!] of training under an experienced master, and, as in the scientific training, the dedicated time does not alone guarantee success. If the dog is successful, however, it will be able to 'repeat the trick.' The repeatability of the trick is, in fact, essential to every canine training and is the very aim of the dog's physical instruction.

Should we conclude from this passage that dog tricks are "parallel" to the mystic's enlightenment?

Another (infantile) syllogism for the "parallel" is self-described in Fig. 7.1. On the left you see some equations from quantum field theory. On the right, there is some text in Sanskrit. Capra knows that an average reader cannot recognize either of the two. He puts them together at the beginning of Part III

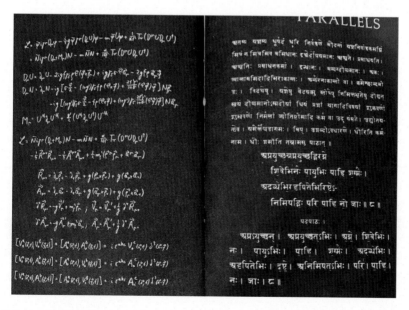

Fig. 7.1 Lack of familiarity with quantum field theory (left) and Sanskrit (right) surely implies the parallel between quantum physics and Eastern mysticism!

of his book entitled "The parallels" to dupe his readers into believing that there is a connection between modern physics and Hinduism. This is an epitome of what I earlier referred to as "proximity implies parallelism" on page 82.

The Tao of Physics and *The Dancing Wu Li Masters* convinced the unsuspecting public that Eastern mysticism had some parallelism with quantum physics to the point that when you have a discussion about Buddhism or Hinduism with a believer these days, they eventually—although hesitant at the beginning—mention quantum physics to convince you of the solidity of their arguments. Those two books may have laid the philosophical foundation of the believers' faithfulness, but the firmness of their conviction is the result of a mechanism that stretches the theosophy expounded in those two books from the abstract mind to the concrete body.

8

The "Quantum" Healer

Few people have distorted and defaced quantum physics more than the mind/body doctor Deepak Chopra. The title of his magnum opus in alternative medicine is *Quantum Healing*. Because of its title, its colossal influence, and the tumultuous history behind its publication, which unearths the professionalism of its author, it is important to examine the path to the appearance of that book.

Mysterious Disappearance of the Maharishi

Maharishi Mahesh Yogi, the founder of transcendental meditation (TM), was undoubtedly one of the most influential figures of the last century, having followers like Beatles and other celebrities. He was also one of the most controversial gurus, not only because of his outlandish assertions such as the claims that advanced practitioners of TM can develop powers of invisibility, mind-reading, perfect health and immortality, but also because of the revelations that his compound in India was the focus of allegations regarding child molestation and death from abuse and neglect.[1]

The hardcover edition of *Quantum Healing* came out in 1989 and its paperback edition in 1990. In the Introduction, Chopra narrates the most remarkable experience in his professional life: "when I was visiting India, one of the greatest living sages had imparted to me some techniques, dating back thousands of years, that he said would restore the mind's healing abilities. I am speaking of Maharishi Mahesh Yogi."

In one of their multiple meetings, the sage told Chopra about some special techniques that Maharishi believed would become the medicine of the future, and asked Chopra to explain, clearly and scientifically, how they work. That is how *Quantum Healing* came to be. From his own words, one gets the unmistakable impression that, were it not for his contacts with Maharishi, Chopra would not have come across the discovery described in his book. In fact, he feels so much indebted to the sage that he dedicates the book "With a full heart and deepest thanks to Maharishi Mahesh Yogi."

As Maharishi's influence on Chopra's discovery is evident throughout the book, one might think that, like an honest scientist, Chopra is acknowledging the conversations he had with the sage and how those conversations might have helped shape his ideas—the way Einstein acknowledged his coworker at Bern patent office, Michele Besso, in his relativity paper and Planck acknowledged Ludwig Boltzmann in his Nobel speech. However, that would be a premature thinking, because the acknowledgement appears only in the printings of the book up to the 14th.

The 16th and subsequent printings are crafted to give the impression that they are the second edition.[2] They contain a single page entitled "Preface to the New Edition." However, the usual practice of imprinting "Second Edition" on the cover is foregone. Furthermore, unlike any ordinary second editions, no preface to the first edition is retained. A conspicuous change in the new edition is that all citations of Maharishi's name are erased. This deletion is especially manifest in the bibliographies of the two printings. In the bibliography of the earlier printings of the book, Chopra writes "I enthusiastically recommend the following eleven books, all of which entered into my own education on these fascinating subjects." Two of those eleven books are by Maharishi. In the bibliography of the 16th and later printings, he also recommends eleven books, but he lists only nine! You can guess which two are missing. In an act that is unbecoming of true scientists and their publisher, and in an ostensibly frenzied rush that can be ascribed only to those who want to hide a damning evidence, Chopra erased all traces of Maharishi's name and the guru's influence on *Quantum Healing*, but he forgot to count the number of the remaining books in its bibliography. What makes all of this suspicious is that there is absolutely no explanation for why the name of Maharishi Mahesh Yogi was removed from the later printings of the book. One can only guess the reason from an article Chopra wrote in *Huffington Post*.[3]

The article details Chopra's intimate relationship with the guru: how he had a vision of the Maharishi lying in a hospital in India, how he flew to India and arranged for the guru to be transferred to a private hospital in London, how he was the only person whose blood type matched the Maharishi's, and how

Maharishi died and after 24–36 hours he was miraculously resurrected. He describes how Maharishi designated him as "the ambassador of TM" with all the prestige and power that came with it, making Chopra a celebrity. Then their relationship soured in July 1993 because the Maharishi suspected that Chopra was sidelining him and taking over the leadership of the transcendental meditation enterprise.

February 13, 2008, when the article appeared in the *Huffington Post*, was just eight days after Maharishi's death and *more than fourteen years* after Chopra broke up with him. What reason could one have for waiting over fourteen years to tell such a self-praising intimate story of a relation between oneself and another person and then publish the story right after the death of that person? Could it be anything but an intention to smear the facts in the absence of the only person who could challenge the validity of the story?

Professional honesty is one of the hallmarks of a good scientist. Chopra is supposedly fulfilling the Maharishi's wish of finding a *scientific* answer to the efficacy of Ayurveda and TM. I hope that the foregoing story gives you a perspective of the viability of arguments that Chopra gives in support of the connection between quantum physics and ancient medical and spiritual practices.

Quantum Oinking

Failure of science-based medicine becomes an opportunity for the promotion of alternative medicine. Science admits to its shortcomings, scripture-based medicine rarely does. After all, how can a practice that has endured thousands of years be wrong? If you follow the ancient protocol exactly as it is written, you will succeed, even in cases where science fails. The only reason that ancient protocols fail is either the incompetence of the practitioner, or the inability of the patient to concentrate deeply on the procedure.

So, what better way to make a science out of an old Eastern scripture than connecting it to a disease for which Western medicine has not (yet) found a cure: cancer. And what better way to inject Ayurveda into the equation than taking cases of cancer in which the patient miraculously recovers from the disease without treatment: spontaneous remissions. The inseparability of Ayurveda and consciousness prompts Chopra to tinker with the hypothetical consciousness of patients experiencing spontaneous remission. How do quantum physics, remission, and consciousness come together? Chopra alleges that the remission of cancer is caused by a "jump to a [higher] level of consciousness that prohibits the existence of cancer."[4] The word "jump" in the context of consciousness takes a special significance:

> The word that comes to mind when a scientist thinks of such [jumps] is quantum. The word denotes a discrete jump from one level of functioning to a higher level – the quantum leap. ... Therefore, I would like to introduce the term quantum healing.[5]

Does Chopra really believe that there actually is a jump in the consciousness of patients experiencing spontaneous remission, even though he cannot measure it? Or does he invent the jump in anticipation of (ab)using quantum mechanics? Had Chopra known that the natural quantum jump is to a *lower* state,[6] perhaps he would have avoided the concept of a "jump." But then again, we have seen how modern gurus ignore reality and truth.

We left Chap. 7 with a couple of syllogisms that were demonstrably preposterous. Through another absurd syllogism, Chopra is trying to convince us that, because of the jump in the—immeasurable, imaginary, unproven, hypothetical—consciousness, spontaneous remission is connected to quantum mechanics. This gives us another opportunity to manifestly establish the absurdity of another New Age "proof." Even though Chopra has no way of measuring the jump in consciousness he does not hesitate to concoct "quantum healing." I can do better. There are ways of measuring the "jump" in many phenomena: Thunder is a jump in the intensity of sound in a storm. Therefore, I would like to introduce the term quantum thundering. Burp is a jump in the level of air released through the mouth, when we eat too much onion or garlic. Therefore, I would like to introduce the term quantum burping. A pig's sound jumps in volume when it oinks. Therefore, I would like to introduce the term quantum oinking. There is as much sense in "quantum healing" as there is in "quantum oinking."

"Quantum" Healer's "Quantum" Theory

Now that quantum physics has become the venue of connecting science with Ayurveda, Chopra has to fabricate his own quantum mechanics. This fabrication uses arrows, straight lines, and curves. If an event A causes an event B, draw a *straight* arrow from A to B. Hitting a billiard ball (A) causes another billiard ball to move (B). Letting go of an apple (A) causes the apple to fall (B). Chopra puts these cause-effect events in one category. This category belongs to the Newtonian world, the world of our experience. What about pairs of events that are not connected by cause and effect?* Chopra creates an ad hoc category to accommodate such pairs of events and accuses quantum mechanics of explaining the relation between those two events. Since those two events cannot be connected by a straight arrow, he bends the arrow in the shape of a "U," puts the bottom half of the "U" in the "quantum world," and draws a horizontal line that cuts through the middle of the "U" to separate the world of our experience from the "quantum world."

How do you explain scientifically the Ayurvedic claim that mind creates matter? Easy! Just connect mind (A) to matter (B) with a U-shaped arrow. But first the connection must become microscopic to be more suitable for quantum physics. Chopra steals the scientific discovery (Ayurveda didn't know of such a discovery) that brain cells synthesize, and release chemicals made up of small chains of amino acids called *neuropeptides*. So, matter reduces to neuropeptides. He then reduces mind to the activity of the brain, thought. However, he turns science on its head by reversing the relational order of the two and declares that thought creates neuropeptides. So, draw a thought (A) and a neuropeptide (B) and connect them with a U—after all, they have to go through the "quantum world." That's it! Chopra not only has fulfilled the wish of his guru, but he has done it using the most exact way available in science.

The following is an application of "quantum Ayurveda" to the feeling of fear and the lifting of a finger. As the attentive reader can see, it is also an epitome of the fallacy in logic called *circular reasoning*.† (I have italicized the sentence containing the fallacy.)

> The mind and the body are both above the line [see Fig. 8.1]. A is a mental event, or thought; all the other letters are physical processes that follow from A. … If you feel afraid (A), then the other letters stand for signals to your adrenal glands, the production of adrenaline, the pounding of your heart, elevated blood pressure, and so on. These are B, C, D, et cetera. All the physical changes that take place in the body can be connected in a logical chain of cause and effect, except for the space after A. *This is the point where the transformation from thought to matter first occurs – and it must occur, or the rest of the events will not happen.*

---

*If the question is too broad, it is intentional. Chopra wants to have a "reason" for the creation of reality by consciousness. He cannot argue that mind or consciousness creates matter as he argues that letting go of an apple causes it to fall. So, he separates mind-matter relation from the Newtonian world and arbitrarily creates a second category. Two events (*any* two events), neither of which is the cause of the other belong to this category.

†An obvious illustrative example of circular reasoning is the following "proof" of the existence of God: God exists, and the sign of his existence is all the animals and plants he created, and he *must* exist because otherwise there wouldn't be any animals and plants.

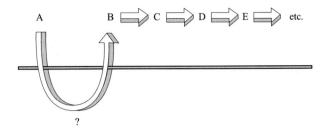

Fig. 8.1 How Chopra's "quantum" mechanics proves the Ayurvedic claim that mind creates matter

> At some point in the lineup, there must be a detour. At that point the lineup breaks down, because mind does not touch matter above the table. If you want to lift your little finger (point A), a physiologist can trace the neurotransmitter (B) that activates an impulse that runs down the axon of the nerve (C), causing a muscle cell to respond (D), resulting in the lifting of your little finger (E). However, nothing a physiologist can describe will get him from A to B – it requires a detour. ... What exactly happens in the ? zone is not known either in physics or in medicine.[7]

The last sentence cleverly whitewashes the entire project of explaining Ayurveda scientifically: I know that A causes B in the quantum domain, but nobody knows what that domain is or how it works.

There is no limit to how absurd pop-spiritualists' reasoning can be. I demonstrated the ludicrousness of the jumping consciousness earlier. Here is another example: I claim that my sneeze caused an earthquake in Mexico City and I can prove it using Chopra's quoted reasoning and Fig. 8.1.

> The sneeze and the earthquake are both above the line. A is the sneeze; all the other letters are physical processes that follow from A. All the physical changes that take place above the earth crust can be connected in a logical chain of cause and effect, except for the space after A. This is the point where the transformation from sneeze to the motion of tectonic plate first occurs – and it must occur, or the rest of the events will not happen.
>
> At some point in the lineup, there must be a detour. At that point the lineup breaks down, because sneeze does not touch tectonic plate above the table. A specific molecule exhaled in my sneeze penetrates the earth crust, sits on a tectonic plate and at an appropriate time activates a seismic wave, which runs to the fault under Mexico City, causing an earthquake there. A seismologist can trace the motion of the tectonic plate (B) that activates a seismic wave that runs

to the fault under the city (C), causing a displacement of the earth crust there (D), resulting in the collapse of a building (E). However, nothing a seismologist can describe will get him from A to B – it requires a detour. ... What exactly happens in the ? zone is not known either in physics or in seismology.

You and the Universe

John Wheeler gave an observer the power of influencing what is being observed through his idea of a participatory universe. But he didn't stop at mere participation. He bestowed upon his observer the power of creation.

Wheeler and two of his students wrote a book entitled *Gravitation*. It is a masterpiece that has taught generations of physicists Einstein's general relativity (GR). Throughout the book, the prosaic Wheeler—to use Dyson's dual characterization of Wheeler as prosaic and poetic—helps explain the intricacies of GR with a peerless combination of clarity and rigor. Then on page 1217, the poetic Wheeler breaks his silence:

> ... may the universe in some strange sense be "brought into being" by the participation of those who participate? ... "Participator" ... strikes down the term "observer" of classical theory, the man who stands safely behind the thick glass wall and watches what goes on without taking part. It can't be done, quantum mechanics says. ... Is this firmly established result the tiny tip of a giant iceberg? Does the universe also derive its meaning from "participation"?

Unsurprisingly, this particular admixture of quantum physics and mysticism is an invaluable gift to New Agers and a passport to the most preposterous claims ever. And the quantum healer takes full advantage of it.

Chopra and one of his colleagues, Menas Kofatos, devote an entire book to the idea of the participatory universe. If the universe is "brought into being by the participation of those who participate," then you and the universe become the same. Thus the title of their book: *You are the Universe*. As is the case with all Chopra's work, the book—regrettably, but predictably—whizzed to the top of the *New York Time*'s bestselling list.[8]

Following the master plan of this book—as outlined in the Preface—I have chosen to dissect *You are the Universe* to expose the emptiness, fallacy, risibility, even hypocrisy that is so common in pop-spiritual works.

Part One of the book, titled *The Ultimate Mysteries*, consists of nine chapters, each headed by a question. The first seven questions are about the universe itself and the last two concern the emergence of life and the working of the brain. By muddling the scientific truth and exploiting the limitation of science regarding the origin of the universe and life, the authors give their baseless theosophical answers to the nine questions. Below is a sample consisting of five questions, the authors' answers, and a scientific rebuttal of the content of those answers.

What Came Before the Big Bang? The answer echoes John Wheeler's *It from Bit*:

> Perhaps something immaterial [information] was the source of orderliness, even if there was only chaos at the physical level during that time. ... This seems like an intriguing thread to follow because other physicists have theorized that when all matter and energy is sucked into a black hole and annihilated, information manages to survive. ... What if information isn't disturbed even under the most extreme physical conditions [of the interior of a black hole]? Perhaps the pre-created state was rich with [non-material] information that was immune to the second law [of thermodynamics] applying at the moment of the big bang.[9]

The recurring words like "perhaps", "seems like", and "what if" heralds the speculative nature of the authors' answer. And their claim of immunity from the fundamental second law of thermodynamics indicates their willingness to discard physics altogether in favor of their nonsensical speculation.

The concocted answer above serves the authors well when they deal with mind and brain, another topic of interest in the book: "By analogy, the information you carry around in your mind can survive all kinds of physical threats. One piece of information is your name. Once you know your name, it doesn't matter if you travel to the steaming tropics or the South Pole, as heat and cold don't cause your name to freeze or boil over." What heat and what cold? Would you remember your name if you were put in a pot of boiling water for one hour? Or stripped naked and tucked under a ton of ice in the South Pole for another couple of hours?

There is absolutely no similarity between black holes and the brain and no connection between the information carried by a physical particle and what we learn and store as memories. There are numerous ways that "the information you carry around in your mind" can get lost. I gave two examples above. But there are more: diseases, age, drugs, and traumas are other examples that can cause the information in your brain "to freeze or boil over."

Analogies could indeed serve as facilitators for understanding abstract concepts, if the limitation of analogies is clearly pointed out. One analogy that we encountered before is quantum "tunneling." Quantum particles penetrate through a—potential—barrier without burrowing, digging, excavating, or tunneling through it. Because the concept is abstract, physicists use the word "tunnel" to describe the process, but they immediately point out the limitation of the analogy by alerting the novice that there are no tunnels or holes in the barrier.[10] The authors are not doing that. They *equate* a black hole—or give the impression of its exact similarity—to a brain and the indestructibility of the physical information to that of the memory. That is plainly wrong.

It is reasonable to assume that the real answer to the question heading the chapter may lie in the unification of Einstein's general theory of relativity (GTR) with quantum physics. After all, big bang, as far as we can tell, is a *microscopic* phenomenon—the domain of quantum physics. GTR, the best language for gravity, works well on large scales. To work on the scale of the big bang, GTR must be consistent with quantum physics, which it is not in its present form. Although physicists have been struggling with this unification for decades, no satisfactory theory has been found. Just as the unification of special relativity with quantum mechanics by Paul Dirac in 1928 led to such exotic ideas as antimatter, the unified theory of quantum physics and GTR, *quantum gravity*, may open completely new vistas and concepts unfamiliar to current generation of physicists. Before that unification, all speculations about what happened before the big bang, are just that, speculations. And since speculations don't cost anything, I propose my own: I stipulate that before the big bang there was a giant dragon who puffed out a fire ball that became the universe. Can Chopra and Kafatos evidentially prove that the giant dragon hypothesis is wrong but their "immaterial" information is right?

Why Does the Universe Fit Together So Perfectly? The question begs the Anthropic Principle.[11] It can even be argued that the authors posed the question to delve into the Anthropic Principle. After all, what can be more in tune with the title of the book than this principle, of which John Wheeler said, "It is not only that man is adapted to the universe. The universe is adapted to man." The authors half-jokingly promote an extremely absurd version of the Anthropic Principle: "The universe came into existence so that I, personally could argue cause-and-effect. ... This might seem like a joke, but if the universe must accommodate human beings, there is no logical reason why it can't accommodate this very moment in time."[12]

Where Did Time Come From? The scientific answer is that time and space were created with the big bang. Current physics cannot explain the mechanism of this creation, nor can it explain the initiation of the big bang itself. The void created by this absence of scientific explanation becomes an opportunity for the authors to impose their groundless mysticism as an alternative answer. After they declare that the quantum domain, and reality itself, has a psychological component and that quantum reality is obedient to the will of the observer, the authors conclude that the best answer to the question is a human answer.

> We didn't have to be present at the big bang for it to have a psychological component. The only version of the big bang anyone will ever know is the story told by human beings … . The same mechanism is producing reality at this very moment. Therefore, the mystery of time exists before our eyes.[13]

We can add yet another example of the absurd syllogisms used by the New Age gurus to our list.[14] I won't bother you with a full competing sentence, all you need to do is to replace "big bang" with any one of the following: boiling of water, melting of ice, onset of a fever, dreams, nightmares, murder of Abel by Cain, rising of the sun, setting of the moon, creation of trees, formation of clouds, … (you can fill the dots with many other choices) in the quote above and note that the resulting statement makes as much (or as little) sense as the original.

What Is the Universe Made of? Because of the equivalence of mass and energy ($E = mc^2$), the components of the universe are usually categorized in terms of their energy content. According to this categorization, it is known that the universe is about 69% dark energy, 26% dark matter, 5% visible matter (mostly hydrogen and helium), and a tiny amount of background radiation. The word "dark" catches the attention of the authors.

> Most of [the universe], 96 percent or so, is 'dark' and therefore unseen and unknown. … as things stand, dark matter and energy, are surmises formulated by painstaking, elaborate lines of reasoning – their actual existence is several steps removed from 'seeing is believing.' Some skeptics warn that physics is flirting with fantasy.[15]

This quote deserves a sharp scrutiny because it conveys a falsehood about physics that can only be diagnosed as an attempt to misinform the reader. And since misinformation has never been as palpable and as dangerous as it is today, the importance of this scrutiny is manifold.

If scientists were to "believe only what they see," we would still be in the Dark Ages. Even the 4% of the universe—the so-called *baryonic*[16] part of it—that Chopra and Kafatos declare as having been "seen," has not been totally captured by the human eye, either directly or through *optical* telescopes. In fact, the vast majority of the baryonic universe has been observed by radio, infrared, ultraviolet, and X-ray telescopes, which capture electromagnetic waves (EMWs) that are "dark" to our eyes. So almost the entire universe is "unseen." But to say that it is "unknown" violates the very nature of science.

One of the best-known entities to mankind is electron. We know its mass and electric charge to nine significant figures—mass of the electron is 0.00⋯0910938356 kilogram (the dots represent 27 zeros) and its charge is 00⋯0160217662 coulombs (the dots represent 15 zeros)—and have used it to provide electricity and operate radios, televisions, computers, smartphones, and electrocardiograms. Our instruments and theoretical models have enabled us to manipulate a small number of electrons to invent transistors and microchips, the crucial elements in computers and smartphones. Yet, no one has ever "seen" an electron. Its existence was a "surmise formulated by painstaking, elaborate lines of reasoning" in 1897 by J. J. Thomson as the negative constituent of atoms, for which "surmise" he won the Nobel Prize in 1906. If Thomson's act was "flirting with fantasy," then so are all great discoveries of physics.

Radio, infrared, ultraviolet, and X-rays are EMWs that are dark to our eyes. So is gravity. And just as invisible EMWs have detectable impressions, so does gravity. When observing the speed of the stars in the outer rim of a galaxy orbiting its center and noting that the gravitational force of the baryonic matter present in and around the center could not account for the speed, Lord Kelvin "surmised," in 1884, that there must be some form of matter incapable of emitting EMWs filling the interior of a galaxy, whose gravity speeds up the stars in the outer layer of the galaxy. As "dark" as it may be, we can "see" dark matter—as it affects the motion of stars at the rim of a galaxy—almost as clearly as (dark) radio waves and electrons.

EMWs were discovered in 1887. Their intense study led to the special theory of relativity (STR) and the notion that EMWs consisted of photons. These two advancements were critical in the unification of STR and quantum physics in 1928. So, more than forty years after the discovery of EMWs the unification of STR and quantum physics became possible. Gravitational waves (GWs) were discovered in 2016. Because of the gravitational nature of dark matter and energy, it is reasonable to assume that GWs—labeled by Chopra and Kafatos as "red herrings"[17]—may play an important role in fully understanding the "dark" objects of the universe. And if historical precedents are of any value, perhaps within half a century we may be in a position to unify gravity and quantum physics and solve the mysteries of dark matter and dark energies.

But history does not progress linearly. It took twenty-two years to observe the EMWs after they were predicted by Maxwell in 1865. GWs were observed one hundred years after their prediction by Einstein in 1916. Assuming that the time interval between discovery and unification is almost twice the interval between prediction and discovery, then we may have to wait another two hundred years before gravity and quantum physics are unified.

These are, of course, wild speculations. Unpredictable social, political, and economic forces play significant roles in the development of science. The scientific environment that existed between 1887 and 1928, out of which came the likes of Planck, Einstein, Bohr, Schrödinger, Heisenberg, Pauli, and Dirac, was created by leading governments which rendered unwavering support for science. The current political climate around the world, especially in the US, does not promise such an environment. On the contrary, the very abundance and popularity of conspiracy theorists, psychics, faith healers, and pop-spiritualists portend a period resembling the Dark Ages.

So, what is the authors' answer to the question, *What Is the Universe Made Of?* Remember John Wheeler's "participatory universe," a.k.a. "experimenter effect?" If atoms and elementary particles do not exist independent of the observer, and if the observer determines the reality of those objects, "then asking what the universe is made of turns out to be the wrong question. ... The universe is made of *what we want it to show us*. ... Can looking at the whole universe, its stars and galaxies, or looking at trees, clouds, and mountains actually transform them? The notion sounds preposterous at this point, but in fact this is the basic claim of the human universe."[18]

Do We Live in a Conscious Universe? The answer should be obvious by now. Nevertheless, Chopra and Kafatos—who in their incessant desire to impress their readers by mentioning the names of professors at prestigious scientific institutions and discussing their ideas only to dispose of them—find an opportunity here to mention MIT. Wheeler's "It from Bit" is reported to have been applied to consciousness by Max Tegmark from MIT: consciousness is nothing but information.[19] Once MIT is mentioned, Chopra and Kafatos discard the notion of information being a substitute for consciousness. What is *their* answer? The other contribution of Wheeler: They argue, "We are participants in reality, which makes us totally involved. Quantum physics is famous for bringing the observer into the whole problem of doing science." Then they play on the readers' most intimate personal feelings to sell their answer: "Reality's message is intimate: 'I have you in my embrace. We are locked together, and the more you try to break away, the tighter my embrace becomes.' ... The observer has nowhere to stand outside reality. ... For human beings, participating in the universe is how we exist. To exist is to be aware.

Astonishingly, the same is true for the universe. Without consciousness, it would vanish in a puff of smoke."[20]

Consciousness: The Cure-All of All Questions

From the preceding discussion, it is not hard to decipher that all answers given by Chopra and Kafatos boil down to one sentence: consciousness explains all the mysteries behind all the questions heading the chapters. In fact, this is made abundantly clear by the authors themselves in the last chapter of the book.[21] I'll reproduce the gist of their answers here.

Mystery 1: What came before the big bang?
Answer: A pre-created state of ***consciousness***, which has no dimension.
Mystery 2: Why does the universe fit together so perfectly?
Answer: It doesn't, because "fitting together" would mean that separate parts would have to be carefully jiggled into place. In fact, the universe is one undivided whole. Its parts, whether we are talking about atoms, galaxies, or forces like gravity, are just qualia—the qualities of ***consciousness***.
Mystery 3: Where did time come from?
Answer: The same place that everything comes from, ***consciousness***.
Mystery 4: What is the universe made of?
Answer: The real building blocks of the universe are qualia, the building blocks of ***consciousness***.
Mystery 5: Is there design in the universe?
Answer: In reality, design is a ***conscious*** perception that is totally malleable.
Mystery 6: Is the quantum world linked to everyday life?
Answer: The quantum domain is another realm of qualia—the building blocks of ***consciousness***—like any other. It needs no link to everyday life because all domains are constructed from ***consciousness***.
Mystery 7: Do we live in a conscious universe?
Answer: Yes. Pure ***consciousness*** gives rise to everything, including the human mind. In that sense, we don't live in a conscious universe the way renters occupy a rental property. We participate—as John Wheeler taught us—in the same ***consciousness*** that *is* the universe.
Mystery 8: How did life first begin?
Answer: As a potential in ***consciousness*** that grew from seed into every variety of living thing.
Mystery 9: Does the brain create the mind?

Answer: No, but the opposite isn't true either—the mind doesn't create the brain. There is no chicken-or-the-egg dilemma, because ***consciousness*** creates opposites all at once.

Consciousness: God in Disguise

Chopra and Kafatos try to distance themselves from traditional religion, because they know that their targeted readers do not believe in the established religions. They mention God frequently, but with a tone of obsolescence and as a concept that needs revision. They resort to a perniciously effective ruse employed by all New Age gurus: inundate the readers with scientific information. Although the information has no relevance to what Chopra and Kafatos eventually want to advertise (e.g., that consciousness is the answer to all questions to which science has currently no answers), the adornment of their pages with scientific truth deceives the readers into believing that science is the basis of the thesis of their book. An in-depth—but straightforward—analysis of *You Are the Universe* reveals the mendacity of physics-consciousness association and confirms the universal arguments for such false association that I alluded to before.*

In the preface of the book, Chopra and Kafatos describe the Vedic sages of ancient India as Einsteins of consciousness whose genius was comparable to the Einstein of the twentieth century and summarize the reality that they preached in four words: "I am the universe." They then bring in today's science and the reality that it presents. Since there cannot be two realities, they argue, the reality of science must agree with that of ancient Indian sages:

> If 'I am the universe' is true, modern science must offer evidence to support it – and it does. Even though *mainstream* science is about external measurements, data, and experiments, which build a model of the physical world rather than the inner world, there are a host of mysteries that measurement, data, and experiments cannot fathom. At the far frontier of time and space, *science must adopt new methods* in order to answer some very basic questions such as 'What came before the big bang?' and 'What is the universe made of?'[Emphasis added]

Several observations reveal the hypocrisy and fallacy of the quote above, and arguably the entire book. First, the premise that "I am the universe" is true is false. It is as true as "God created the universe in six days." Why

---

*See page 22.

should anyone accept what some ancient Indian sages said thousands of years ago? Were they somehow more scientific than the ancient Hebrew sages, or ancient Mayan sages, or ancient Greek, Egyptian, Babylonian, Persian, … sages? Second, science *does not* offer any evidence to support the falsehood, "I am the universe." This claim of support is injected to deceive the reader to believe that the reality of Vedic sages is the same as scientific reality. In fact, the very next sentence negates the claim and the "Even though" at the beginning signals the negation. Third, the qualifier "mainstream" indicates the authors' rejection of science as a whole. What they are referring to are fringe scientists or qualified scientists, such as John Wheeler, whose science (of "the physical world") has nothing to do with their mystical beliefs (in "the inner world"). Finally, to instruct science to "adopt new methods" to answer some very basic questions such as "What came before the big bang?" and "What is the universe made of?" is tantamount to rejecting scientific reality altogether and replacing it with the reality of Vedic sages. And this becomes amply clear when Chopra and Kafatos answer all the posed questions by invoking consciousness at the end of their book.

How is their doctrine different from religion, which they dismiss at the outset? To find out, simply replace all the occurrences of "consciousness" in their answers discussed above with "God" and note that the new answers are as (in)valid—and arbitrary—as theirs. As God is the answer to all phenomena in a traditional religion, so is consciousness in the religion of Chopra and Kafatos. The only difference between the two is that the former had the darkness of the Middle Ages to keep the public ignorant and the political might to stifle any opposition, while the latter, incapable of the tactics of the former, resorts to the illiteracy of the public and the profit-seeking media corporations exploiting that illiteracy.

Because pop-spiritualists and pseudoscientists insist on using it so profusely, a few remarks about the word "mainstream" is in order. While in political, social, journalistic, … circumstances—because of the varieties of social "theories"—it makes sense to separate the largest group of practitioners and christen them "mainstream," no such designation is appropriate for science.

The word "mainstream" in science is concocted by pseudoscientists to legitimize their own nonsense. Scientists, according to them, are divided into two distinct categories: mainstream (consisting of ordinary scientists) and revolutionaries (like Copernicus, Galileo, Newton, and Einstein), and they put themselves in the second category because they know and everyone else knows that they don't belong to the first category. This categorization is completely false. There is only one science and only one category of scientists. If we were coerced to categorize scientists, then we could identify three categories:

1. Those who do mainstream science.
2. Those *mainstreamers* who bend the mainstream.
3. Those who leave the mainstream and turn into crackpots.

The overwhelming majority of scientists belong to the first category. Scientists like Galileo, Newton, Dalton, Crick and Watson, Planck, and Einstein belong to the second category. People in the third category may once have been accomplished scientists in the first category; however, for various reasons, they left the mainstream science, and with it, science itself. People like Deepak Chopra, Andrew Weil, and Fritjof Capra by their own admission, are no longer mainstream scientists. And certainly, they don't belong to the second category![22]

Quackia: Quanta of Fairies

Space limitation dissuades me from dissecting the second part of *You are the Universe* in as much detail as I did Part One. However, the authors' use of the word "qualia" in their answers prompts some scrutiny. Qualia is Chopra's and Kafatos's way of deceitfully presenting consciousness as a science on a par with quantum physics. They compare qualia with quanta and claim that

> The concept [of qualia] is tremendously important, even though the average person has never heard of it. With qualia you can change your perception – or not. With qualia you can alter reality – or not. Qualia refers to how we experience life rather than how we measure it. The word *qualia*, which is Latin for 'qualities,' is a tag for a world that is as far-reaching as quantum physics, but points in the opposite direction, away from physical objects and toward subjective experience. Whereas quanta are 'packets' of energy, qualia are the everyday qualities of existence – light, sound, color, shape, texture – whose revolutionary implications we've already begun to describe.[23]

Are qualia as far reaching as quanta? How did quanta come to be? How did qualia come to be? How many scientists were involved in the creation of quanta? Were there *any* scientists in the creation of qualia?

Gustav Kirchhoff, a German physicist, analyzed the dependence of the color of the glow of a heated object on temperature in the 1850s. He also investigated the intensity of various wavelengths in the glow for a given temperature. His efforts led to the proposal of a mathematical formula—dependent on both temperature and wavelength—whose curve as a function of wavelength at a fixed temperature has become known as the *black body radiation curve* (BBRC).

In the last decade of the nineteenth century, one of the most active areas of research was finding the mathematical formula describing the BBRC. Wilhelm Wien, the leading German researcher, found a formula for BBRC which seemed to agree well with the observations done at the time. Max Planck found an alternate way of deriving Wien's formula and submitted his derivation to *Annalen der Physik* in November 1899. As the derivation of Planck was being printed, there was already experimental indications that Wien's formula was not quite right. After scrutinizing his formula, Wien concluded that the formula must be valid only for short wavelengths and not necessarily for long wavelengths.

On October 7, 1900, Rubens, one of Planck's experimental colleagues, visited Planck and reported his (Rubens') recent findings concerning the long-wavelength behavior of black body radiation. He also mentioned the agreement between his observations with a formula derived by the British physicist Lord John Rayleigh in June. Upon receiving this news, Planck immediately set out to generalize Wien's formula so that it agreed with short and long wavelengths of the black body spectrum. One of Planck's students has reported on these historic events:

> The same evening Planck reported this [new] formula to Rubens on a postcard, which the latter received the following morning. One or two days later Rubens again went to Planck and was able to bring him the news that the new formula agreed perfectly with his observations.

Planck presented his new formula under the modest title *An Improvement of Wien's Spectral Law* to the German Physical Society on Friday, October 12, 1900. All the subsequent observations and precise measurements, up to the present, point irrefutably to the correctness of Planck's black body radiation formula.

One particular observation is of significance. Figure 8.2 shows the plot sent by <u>C</u>osmic <u>B</u>ackground <u>E</u>xplorer—an observatory, dubbed COBE, mounted on a satellite—and presented in the winter meeting of the American Astronomical Society, held outside Washington DC, on January 13, 1990. The data points are measured intensity of the Cosmic Microwave Background as a function of frequency. As predicted decades earlier, the figure proved that CMB was a perfect black-body radiator, and its BBRC indicated a temperature of 2.725 degrees Kelvin.

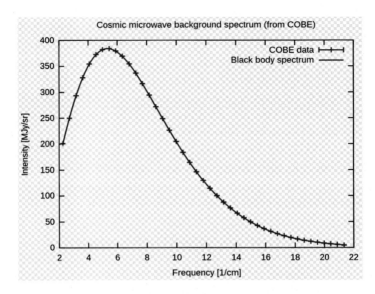

Fig. 8.2 COBE satellite showed that Cosmic Microwave Background is a perfect black body radiator

To explain the BBRC, Planck had to assume that EMWs are made up of bundles of energy that he called *quanta*. He reached this assumption in mid-November of 1900 but presented it publicly at the meeting of the German Physical Society in Berlin on December 14, 1900. On this day the quantum theory was born.

The universe itself confirms the BBR curve and the idea of quanta. Do Chopra and Kafatos have anything to back up their qualia? Do they have any evidence even for the universal consciousness, of which qualia are supposedly elemental units?

Subsequent to nearly half a century of painstaking effort by dozens of top-notch physicists, a genius like Planck, "after some weeks of the most strenuous work" of his life,* *was forced* by nature itself to propose that electromagnetic radiation consisted of packets of energy, or quanta. Chopra and Kafatos, with no effort or supporting evidence whatsoever, throw in the idea that soul, spirit, cosmic energy, pneuma, or consciousness consist of qualia, and—perhaps because "qualia" sounds like "quanta"—call that as revolutionary as Planck's quanta.

---

*See page 43.

Since qualia have no evidential support, I would like to throw in my own equally non-evidential idea of *quackia* and claim that

> quackia refers to how we experience life rather than how we measure it. Quackia is a tag for a world that is as far-reaching as quantum physics, but points in the opposite direction, away from physical objects and toward subjective experience. Whereas quanta are 'packets' of energy, quackia are the everyday qualities of existence—insights, thoughts, dreams, imagination. And just as electromagnetic radiation consists of quanta and consciousness consists of qualia, so do ghosts, goblins, ghouls, Santa Claus and leprechauns – being manifestations of imagination – consist of quackia.

Claims without evidence are opinions. They abound in philosophy, politics, religion, ethics, mythology, and pop-spirituality. In science, they are discarded. It is true that breakthroughs in science start with claims that are initially non-evidentiary. That we call them "breakthroughs" speaks to the fact that, in the end, countless pieces of evidence certify their validity. However, claims will eventually turn into opinion—or cult, like bootstrap hypothesis or cold fusion—if no evidence shows up.

There was a mountain of evidence for the idea of quanta at the time of its proposal, not the least of which was the explanation of the black body radiation. More importantly, in subsequent years, it prompted numerous experimental and theoretical investigations leading up to the quantum physics of 1926–1928, which, in conjunction with relativity, has become the foundation of our knowledge of the universe, large and small.

What is the evidence for qualia? Chopra and Kafatos: "So what kind of evidence would satisfy the everyday rational person [read "our gullible readers"] (we'll exclude die-hard skeptics [read "scientists"], who are beyond persuasion) that the universe is conscious?" They start by telling their readers the falsehood that "In cosmology, there are basically two camps, 'matter first' and 'mind first' " This is a concocted statement with no purpose other than to mislead and misguide the unsuspecting reader. There is no cosmologist whose science—not his philosophy, theosophy, opinion, etc., of the like that prompted Bohr, Heisenberg, Schrödinger, Pauli, Wheeler, and Dyson to make mystical utterances—starts with the mind.

For their evidence, Chopra and Kafatos pose more questions:

> Was the infant cosmos pushed into existence by physical forces or by a mind? Is it enough to have bricks without a bricklayer? ... We need a bricklayer who functions for science the way God functions for religion. The universe has

infinitely more complex building blocks than a cathedral, and the only candidate for a bricklayer who can keep them all straight is the cosmic mind.[24]

The last two sentences of the quote embody a bizarre syllogism which seems to associate God with a cathedral and the cosmic mind with the universe. Do the Torah, Bible, and Koran confine the might of the almighty to the construction of temples, cathedrals, and mosques? Isn't it a blasphemy in those religions to say that God is not mighty enough to have created the universe because "universe has infinitely more complex building blocks than a cathedral?" Science and religion both try to explain the universe. The difference is that science does not need a bricklayer; religion does. The invocation of a bricklayer for the universe is a sign of religiosity whether you call the bricklayer God or cosmic mind.

Here is the gist of Chopra's and Kafatos's evidence: Physics cannot explain the very occurrence of the big bang and how it led to the complex structures in the universe based on the known physical theories. This lack of explanation is therefore *evidence* for a mind or consciousness—whose building blocks are qualia—that created the cosmos.

And here is my evidence for quackia: Everyone has heard of ghosts, goblins, ghouls, Santa Claus, and leprechauns, and some even claim to have seen them. These entities are as old as civilization itself. Denying them is denying our civilization. Despite their prevalence in our mind, scientists cannot explain them based on building blocks of matter such as atoms and molecules. It is therefore "evident" that a different kind of building block is needed to explain these entities. This building block is quackia.

Actually, I have a much stronger evidence than Chopra and Kafatos. My evidence—shown in Fig. 8.3—*proves* the parallel between goblins, ghouls, Santa Claus, leprechauns, … and quantum physics. The proof has the same compelling validity as Capra's proof of the parallel between Eastern mysticism and quantum physics captured on page 95.

"Quantum" Healer Attacks Physics

Chopra has been called the prophet of alternative medicine.[25] His marriage of quantum physics with Ayurveda, meditation, and consciousness—together with his celebrity status, which gives him free access to powerful microphones—has had a detrimental effect on science and rational thinking. His direct attack on physics is particularly damaging.

8 The "Quantum" Healer

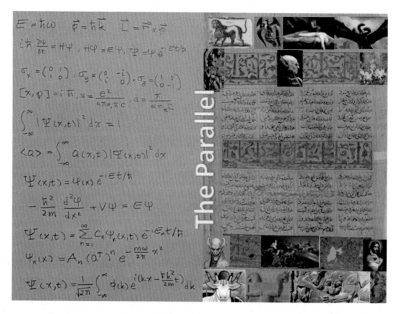

Fig. 8.3 A strong evidence that there is a parallel between quantum physics and fairies, ghosts, goblins, and ghouls

On February 29, 2016 Deepak Chopra, Menas Kafatos, and Rudolph Tanzi wrote an article on *HuffPost* entitled "Why Gravitational Waves Are Red Herrings."[26] The authors dismiss the significance of the discovery of the gravitational waves (GWs) as simply the fulfillment of "a prediction that was almost a century old." If dismissing discoveries because they are old predictions had historical precedent, then we should have had to dismiss the discovery of electromagnetic waves (EMWs) because they were discovered 22 years after they were predicted. Imagine where we would be today without EMWs: we would have no radios, no televisions, no computers, and no smartphones. We would have to travel on carriages because many parts of automobiles require EMWs for either their production or their operation. In short, without EMWs, our civilization would be pushed back to the nineteenth century.

The authors' dismissal of the discovery of GWs takes a bizarre turn when they proclaim that GWs "don't imply anything for quantum mechanics." This statement boggles any rational mind. Why should the discovery of GWs imply anything else at all? If a cure is found for cancer, are we to dismiss it because it doesn't "imply anything for" heart disease? Would Chopra, Kafatos, and Tanzi have declared EMWs as insignificant in 1887 because they did not imply anything about, say, atoms?

But what is most troubling is the audacity with which the authors attack science directly. And in their assault, they employ the age-old tactic of pseudoscience: exploit the scientific achievements until you reach the frontier of research where unanswered questions are being investigated; then discard science as a futile exercise because of its limitation and inject pseudoscientific alternatives as answers. In the last paragraph of their article, the authors exploit the idea of the four-dimensional space-time discovered in relativity theory and then hold physics hostage until

> "We figure out how the dark, watery, mush of brain tissue produces the image of a four-dimensional world when it has no light or sound inside it."

To be clear, this expectation of the authors will never be fulfilled because there exists no *image* of a four-dimensional world. No one, not even Einstein himself, could visualize the four-dimensional space-time of relativity. Space-time is not even a straightforward generalization of the three-dimensional space, because time, as the fourth dimension, has a completely different property than the other three dimensions.* The four-dimensional space-time is a convenient mathematical construct that enters the equations of relativity and facilitates the understanding of the theory.

The brain is an indispensable tool in the hands of the authors, not just because they demand of it the impossible visualization of the four-dimensional space-time, but also because it is the source of consciousness. So, the hostage takers have a second demand: we must

> "determine once and for all where consciousness comes from."

Until these two demands are met, "gravitational waves are just another set of data to pile on to the mountain of data being churned out every day. Although fascinating and a great advance in science itself, they are red herrings as far as understanding reality is ultimately concerned."

Without any pretense of knowing the answer to the authors' second demand, I'll examine what a *scientific* investigation of the source of consciousness might look like.

---

*For example, while you can easily go to the right and return to the left, move forward or backward, and go up and come back down, you can never go back in time or travel to the future.

Where Does Consciousness Come From?

Before answering the question, let me pose three related questions whose scientific answers we already know: What is motion? What is energy? What is life?

Aristotle tries to answer the first question *holistically* in his *Physics*. The tome is composed of eight Books. In Books I–IV, Aristotle spends pages upon pages defining motion (which, according to him, includes the healing of the wound, the heating of the cold, the lighting of the dark, and, of course, the galloping of a horse), making philosophical arguments against his opponents' views on motion, defending his interpretation of motion, classifying motion, disproving the physical existence of infinite, and a long philosophical discussion of space, time, and the void. Books V and VI are further elaboration on the classification of motion and rest. It is not until the end of Book VII—and thus very near the end of *Physics*—that we encounter a real discussion of motion, the kind that should be presented at the very beginning: movement of real objects. Here Aristotle tries to be slightly less general and more quantitative. He argues, in the purest philosophical connotation of the word, that the speed of an object is proportional to the force acting on that object. This (law) is entirely speculative, and, indeed, wrong. Had Aristotle, ever so informally, tried his hypothesis on a real moving object, he would have immediately caught the error in his equation. He would have discovered that if a force (such as the push of one man) acts on an object and moves it on a floor with constant velocity, then the addition of an equal force (such as the push of an additional man) does not cause the velocity of the object to double instantaneously as the law suggests, but accelerates the object slowly, not just to twice the velocity, but to greater and greater speed.

Galileo gave the scientific—and necessarily *reductionist*—answer to the question by noting that motion is a *property of material objects* described by such *measurable* quantities as displacement, velocity, and acceleration. By *reducing* his study of these quantities to the motion of a ball rolling down an inclined plane, Galileo discovered the *universal* first law of motion, which holds even today more than four centuries after its discovery.

The word "energy" has been particularly deformed in the New Age literature: it is that non-material thing that a medium feels when the dead talk to him; it is that power that is transferred to the patient when a psychic healer brings her hand close to the patient's body; it is the force that you experience when you go into a deep meditative trance. Scientific energy is none of that. Scientific energy, like motion, is a *measurable property of material objects* that comes in a variety of forms. No energy exists that is not carried by a material object.[27]

Now to the third question. Life has a unique characteristic that is absent in nonliving things. The first major categorization of nature, known even to prehistoric humans, was the division of objects into animate and inanimate. Bury a bean, which looks as inanimate as a pebble, in the ground and after a few weeks a green leaf appears on the burial point. Do the same with any other bean, a—perhaps different—green leaf appears. Do it with a pebble. Nothing will happen. Living things seem to possess something that nonliving things do not; a soul, a spirit, a pneuma, a *vital force* that gives purpose to animate objects.

Aristotle was a *vitalist*, and as in all other scientific disciplines, dominated the thinking of biologists for a long time. However, unlike his dominance in physics, which was toppled in the sixteenth century, his authority in biology proved to be much more resilient. The reason for the stubborn persistence of vitalism was biology's almost infinite diversity and the unimaginable complexity of even the simplest living thing. By the end of the nineteenth century, chemistry was dealing with atoms and molecules as the fundamental building blocks of matter interacting among themselves via the electromagnetic force, and physics was beginning to look inside the atom itself. The diversity and complexity of the inanimate world was thus reduced to a few dozen atoms interacting through the electromagnetic force and combining to form molecules of all terrestrial objects under consideration.[28]

No such reduction seemed to be operative in biology. Although the cell theory of living things was firmly established in that period and the attention of biologists had shifted to the study of this building block of biology, the diversity and complexity of cells were still of such magnitude as to render their study enormously cumbersome. It wasn't until the middle of the twentieth century when the structure and function of the molecule of life, DNA, was unearthed, that biology became a science on par with physics and chemistry.

Nevertheless, there were—and still are—biologists who speak of a genetic program possessed by all organisms and conclude, "Nothing comparable to it exists in the inanimate world … the possession of a genetic program provides for an absolute difference between organisms and inanimate matter."[29] This strong (molecular) differentiation between animate and inanimate objects is nothing but vitalism at the molecular level. Molecular biologists do talk of "genetic program." However, they do not accept the "absolute difference between organisms and inanimate matter." Their aim is to explain that "genetic program" in terms of physical and chemical laws, just like the inanimate matter.

Biology has made unprecedented progress since the discovery of the double-helical nature of the DNA, and the "molecular revolution" has had its unexpectedly colossal impact on our lives. No one, not even the most prophetic visionaries, could have thought in the 1950s that DNA would someday be used to discover that the entire population of Europe is the descendants of only seven women; that DNA testing could prove without any doubt the guilt or innocence of a suspect; that bones lying under a parking lot in Leicester belonged to King Richard III of England; that DNA could be engineered to manufacture drugs, new plants and crops capable of resisting diseases and feeding large population of people who otherwise die of malnutrition. But these are merely *applications* of molecular biology. What is equally important is that we have begun to understand what life is. The fundamental question confronting molecular biologists now is how billions of only five kinds of atoms—atoms identical to those found in the inanimate objects—can get together and form a molecule that is capable of duplicating itself.[30]

So, the scientific answer to the question, "What is life?" becomes: life is a property of the DNA molecule and to unravel life, we first need to understand DNA. Once that is accomplished, we must fathom how the DNAs of different cells in different organs of a living being interact to make that being what it is. We are a long way from such understanding.

With the preceding introduction, the original topic of consciousness becomes more tangible. How do we scientifically answer the question "What is consciousness?" Just like motion, energy, and life, the question begs a second question: What is the material body whose property, manifestation, attribute is consciousness? The obvious answer is our brain.

Reductionism, the process of studying building blocks of matter, is detested by Eastern mystics, holistic New Agers, and some biologists. It has helped us understand the tiniest particles and the largest galaxies, even the universe itself. Reductionism also gave us the cell theory of living things, the importance of the nucleus of a cell, the chromosomes, and eventually, the molecule of life, DNA. Whenever reductionism was used, science progressed, and when the discovery of the building block was hindered, so was the advancement of science. A prime example of this hindrance, and the power of the reductionism that overcame it, is neurology. And if we assume that consciousness is the manifestation of the *material* brain, then the reductive study of neurology may give us a clue to the question, "What is consciousness?"

Although the cell theory of living things was established decades earlier, most neuroanatomists in the late nineteenth century thought that the brain was an exception; that it was not composed of cells. Even under a microscope, the brain looked like a tangled morass (a reticulum) with no apparent cellular structure. The resilience of this holistic reticular view is clearly demonstrated in the works of the Italian anatomist Camillo Golgi, who, toward the end of the nineteenth century developed a method of highlighting the morphology of just a few neurons (the brain cells) in any particular region of the brain. Golgi's method hardly ever worked, but paradoxically, precisely because of its frequent failure, isolated neurons could be stained and seen under an optical microscope. Despite his crucial discovery, Golgi remained an ardent subscriber to the 'reticular theory,' which held that the brain contained no discrete component.

The cell theory of the brain is attributed to the Spanish neuroanatomist Santiago Ramon y Cajal, who, with the help of Golgi's method, proposed that neurons are actual cells, and the activity of the brain is a unidirectional transmission of electric impulses between them. Cajal's proposal initiated an explosion in neurology and marks a milepost in modern neuroscience and the fundamental understanding of how physical and chemical processes initiated in neurons and transmitted to neighboring neurons culminate in a thought. The brain was the last stronghold of anti-reductionism, and with the triumph of the cell theory of the brain, the molecular vision of life reached its anticipated consummation. Golgi and Cajal shared the Nobel Prize for Physiology or Medicine in 1906.

Today's science explains many outstanding puzzles by reducing objects to their elemental parts.[31] The reduction of the hydrogen atom to a negative electron and a positive nucleus held together by electrical forces was not only necessary for understanding the atom in 1926, but for paving the way for the discovery of Schrödinger equation, which made that understanding possible. Why some substances conduct electricity and others don't has been explained only through a quantum mechanical study of their atomic structure and how they arrange themselves. Without this study, we cannot explain why silicon is an insulator when it is cold and a conductor when it is heated up properly. Superconductivity—the property of some materials whose electrical resistance disappears at extremely low temperatures—was discovered in 1911. It remained unexplained until 1957 when the investigation of the property was *reduced* to the modification of the electrical interaction of electrons in a solid due to the presence of the atoms making up that solid. Similarly, the discovery of neurons helped to quantify intelligence: in general, the larger the number of neurons in the brain of a species, the more intelligent that species. A worm has only about 300 neurons in its brain; an ant approximately 250,000; a honey bee around one million; a pigeon about 300 million.[32]

Neuroscientists have discovered the elemental activity of the brain to be electrochemical impulses passed on from one neuron to the neighboring ones.

From this intricate communication between brain cells arises our intelligence, our mind, and our consciousness. The exact mechanism of how this happens is not known, or as New Age gurus would say, "is a mystery." But hydrogen atom was a mystery before 1926 and superconductivity was a mystery before 1957, and now they are no longer mysteries. The "mystery" of intelligence (or consciousness) is several giant steps away from the question of how a single neuron communicates with its neighbors. Neuroanatomists are trying to answer this simpler question first. The question of consciousness ought to await a reasonable understanding of neurons, just as the question of superconductivity had to wait until the puzzle of the hydrogen atom was solved.

But New Age gurus can't wait for a scientific answer to the origin of consciousness, which will eventually arrive. In fact, they are actually delighted to exploit this unknown territory of science to inject their own unsubstantiated, unproven, baseless hypothesis as an alternative. Vitalism aimed at separating biology from physics and chemistry and giving it a unique autonomous status in science. Consciousness is the new face of vitalism among New Age gurus, whose goal is to separate it from the physical and chemical brain and give it a unique autonomous status, thus replacing the reductive investigation of the brain with holistic mantras, making you believe that your consciousness and the Vedic universal consciousness are one and the same.

A Modern St. Augustine

One of the factors that solidified Christianity after the fall of the Roman Empire was putting an intellectual mask on the teaching of the Bible. As Rome itself did not produce many thinkers, Greek philosophy dominated the intellectual circles of Europe after the decay of Rome. St. Augustine of Hippo had a prominent role in establishing a common ground between Bible and Plato's philosophy, thus putting an intellectual face on Christianity and legitimizing it as a religion founded on reason and intellect.

Almost sixteen centuries later, New Age gurus are doing with Eastern theology and quantum physics what the intellectuals of Antiquity did with Christianity and Platonic philosophy. If St. Augustine of Hippo was a luminary of Plato-Christianity connection, Deepak Chopra is the giant of quantum-physics-Eastern-theology association.

Chopra's impact on the collective thinking of the West today is comparable to the impact of St. Augustine of Hippo—whose *Confessions* is arguably the theoretical manifesto of the Dark Ages—on the collective mind of Europe sixteen centuries ago. And because the force behind Chopra's influence is not Inquisition-like torture but the preponderance of scientific ignorance around the world, his effect is manifold: The will of the scientifically challenged masses gives a democratic power to nonsense that far outweighs the autocratic power of the Church in the Dark Ages.

The similarity between Chopra and St. Augustine is striking when it comes to their attitude toward science. In 400, St. Augustine said,

> There is another form of temptation even more fraught with danger. This is the disease of curiosity … It is this which drives us on to try to discover the secrets of nature, those secrets which are beyond our understanding, which can avail us nothing and which men should not wish to learn.[33]

Chopra's attack on science has been increasingly manifest in recent years. His "red herring" remark on the gravitational waves is not an isolated incident. As if parroting St. Augustine's quote above, in 2013, Chopra said,

> Human nervous system is not necessarily the most developed organ in evolution to give us a clue to the mysteries of life. Evolution was there for the survival of the species *not for uncovering truth*.[34]

The teachings of St. Augustine of Hippo helped solidify the authority of Bible as an alternative to science, portrayed as "the disease of curiosity." Then we fell into the abyss of the Dark Ages. Chopra is teaching us that our evolution was only for our survival, and "not for uncovering truth." And since science is the ultimate tool for uncovering truth, we should abandon science, and instead—although he doesn't say it explicitly—we should rely on the Vedic sages of ancient India and sing the mantra, "I am the universe."

Where will this advice take us? No, where has this advice taken us already?

9

Basic Building Blocks

"Holistic approach" has become a trendy expression among New Agers and pop-spiritualists. Perhaps because of the homonymy of the first four letters of the phrase and the revered word "holy," anything *holistic* has instilled in the collective mind of the public a sense of reverence matched only by religious devotion. Some people experience an almost divine connection with holistic medicine, holistic treatment, holistic clinic, and holistic food.

Science Cries for Reductionism

The antipode of holistic method, reductionism, has been indispensable in the growth of science in all human history. The forefathers of Egyptian priests sought the source of catastrophic happenings on Earth, first in a few objects in the sky and then in the Sun, which was thought to be the seat of the supreme god, Ra. *Reducing* the number in the source significantly eased dealing with it. This was in contrast to the monks in the Far East who meandered in the crowd of spirits on Earth in search of a holistic answer. As a result, the former started astronomy and handed it over to the Greeks who lifted us to a scientific glory unmatched in history before Renaissance. The latter, on the other hand, missed the opportunity of the most important part of science, theorizing, thus limiting itself to inventing firecracker and finding some practical application of magnets in navigation. Without the Egyptian struggle to understand God, we would not have had Greek science, as antithetical as the two may seem.

In ancient philosophy, reductionism—combined with materialism—led to some remarkable findings more than twenty-three centuries ago. The Greek atomic philosophy reduced everything to material atoms, and by doing so made some extraordinary achievements akin to modern science. Why are the hands of the bronze statues at the gates of cities shiny while the rest is covered with green rust? Because the city visitors pick up the atoms of the green rust on the surface as they shake hands with the statues before entering the city. Where do the clouds come from? As the heat of the Sun impinges on the surface of the oceans, water atoms are released to the sky to come together and form clouds. What is a sound? It is the atoms sent by the source interacting with our ears. How do we smell? A flower emits atoms and the sensors in our nose interact with those atoms.

In Greek astronomy, reductionism led to heliocentrism. Rather than simply cataloguing innumerable celestial bodies, Aristarchus of Samos *reduced* the objects of his study to Earth, Moon, and Sun. From the shadow of Earth on Moon in a total lunar eclipse, Aristarchus could estimate the diameter of Moon to be a third of Earth's diameter. He then calculated the Earth-Moon distance by measuring the angular size of the Moon and estimated this distance to be about 25 Earth diameters.* For the Earth-Sun distance, he obtained a value that was roughly twenty times Earth-Moon distance. Since a full Moon and the Sun at high noon appear to be about the same size, Aristarchus concluded that Sun must be twenty times bigger than Moon or about seven times bigger than Earth. He then argued that a smaller object is more likely to go around a bigger object rather than vice versa. He thus came up with the heliocentric model of the solar system almost 1800 years before Copernicus. Aristarchus would not have reached this conclusion without his reductive approach.

With the exception of some contribution by Archimedes, physics started with Galileo in the sixteenth century ACE. The dominant Aristotelian physics of the time was a holistic treatment of nature whose definition of motion included not just the galloping of a horse, but also the healing of the wound, the heating of the cold, the lighting of the dark, the budding of a tree, and all the infinite varieties of *change*. Galileo's emphasis on experimentation, by its very nature, is reductionist: one can experiment with only a limited number of objects. In his study of motion, for example, Galileo focused on the sliding of a block on an inclined plane. This specific, restricted experiment led him to the discovery of the universal *first law of motion*.

Newton generalized the first law, defined what force is in the study of motion, and discovered the *second law of motion*. Furthermore, by *reducing*

---

*The interested reader can find more details about calculations of sizes and distances of Earth, Moon and Sun, including some helpful figures, in Appendix.

the objects of his inquiry to the Moon and an apple, he came up with the mathematical formulation of the *Universal Law of Gravity*, which, when combined with the second law, predicted the motion of the planets around the Sun.

Any progress made in science—not just physics—has relied on reductionism. Chemistry was revolutionized by Dalton's revival of the Greek concept of atoms, at which point chemistry became the study of molecules formed by atoms. Biology underwent enormous advances once the cell theory of living things was firmly established and its breakthroughs exploded when the double-helical DNA, a monstrous molecule, was discovered.

The study of Dalton's atoms themselves fell on the shoulders of physicists. In 1911, Rutherford used alpha particles to probe the interior of the gold atom.[1] As Rutherford tried to verify a kind of "holistic" atom, where the negative and positive charges were mingled in a whole, he was shocked to find that the positive charge was completely detached from the negative charge and concentrated in a region that is one hundred thousand times smaller than the atom itself. Rutherford called this heavy concentration of positive charge *nucleus*. If nucleus were a marble, it would be sitting at the center of an atom more than a kilometer in radius.

Ordinary Matter

Schrödinger discovered his eponymous equation while studying hydrogen atom. His approach was mathematically reductionist, meaning that his equation—which happens to be a mathematical abstraction of the conservation of energy—as a *whole was the sum of its parts.*

Fundamentally, there are two kinds of energy: kinetic (KE) and potential (PE). KE is associated with a single particle, while PE is a manifestation of the force *between a pair* of particles. So, we speak of the electrical potential energy between a pair of charged particles. Therefore, for hydrogen atom, Schrödinger equation is the sum of three terms: KE of the nucleus, KE of the electron, and PE of electron-nucleus pair. This equation can be solved analytically (exactly). For helium, with one nucleus and two electrons, Schrödinger equation is the sum of six terms, three KEs and three PEs.[2] Reductionist application of the Schrödinger equation to ordinary (non-relativistic) matter has been extremely successful.

Atomic Physics: In 1929 Norwegian physicist Egil Hylleraas applied the Schrödinger equation to the helium atom. This is a three-body problem, which cannot be solved exactly, even in classical physics. However, there are methods of approximation that can calculate various properties of a three-body system fairly accurately. Hylleraas's approximation gave values for the energy levels of helium that were quite accurate. Hylleraas's work on helium opened the way for the application of Schrödinger equation to other atoms.

Astrophysics: Nuclear fusion is the primary source of stellar energy. In a multi-stage process, four hydrogen nuclei (protons) at the core of a star fuse together to form a helium nucleus (two protons and two neutrons) and other particles, including photons, carrying a tremendous amount of energy. In the first stage, two protons fuse together to form a deuteron[3] plus some very energetic particles. Classical physics cannot explain this process because the closer the protons get to each other, the larger their electrical repulsion; at contact, the repulsion is almost infinite. This is called the *Coulomb potential barrier*. Quantum theory allows the *tunneling*[4] of one proton through the potential barrier and subsequently sticking to the other proton via the strong nuclear force. The photons produced at the core travel millions of years (yes, millions of *years*, due to the enormous density of the interior of stars) to reach the surface of the star and let it shine.[5]

Chemistry: In 1927, Walter Heitler and Fritz London applied the Schrödinger equation to the diatomic hydrogen molecule. The molecule consists of two nuclei and two electrons. The equation is therefore the sum of ten terms: four KEs and six PEs. Although some simplifying assumptions were used to reduce the number of terms in the Schrödinger equation, the equation could be solved only approximately. The solution was related to the phenomenon of the chemical bond. Heitler-London method was generalized by other physicists and chemists and led to the *valence bond* method of studying chemicals. An alternative approach to applying the Schrödinger equation to molecules led to the *molecular orbital* method. Today, chemists employ ingenious methods of approximations to understand molecules of various sizes and shapes using the Schrödinger equation.

Condensed Matter: Emboldened by the success of the Schrödinger equation in simple systems, physicists applied it to various solids, liquids, and gases. The field is now called *condensed matter physics*. The reductive approach adds up all the KEs and PEs of the (practically infinite number of) particles in a solid, or liquid, or gas and seeks a solution to the resulting Schrödinger equation. Surprisingly—in the case of solids at least—this infinitely long Schrödinger equation can be solved approximately with remarkable precision, mostly because the large number of particles permits certain simplifying symmetry assumptions.[6]

By applying the Schrödinger equation to solids, liquids, and gases, condensed matter physicists have been able to explain why solids are electrical conductors, insulators, or *semiconductors*. The latter—an example of which is silicon—act like conductors at high temperature and insulators and low temperature. And why some solids completely lose their electrical resistance at certain very low temperatures and become *superconductors*. Other byproducts of the reductionist application of the Schrödinger equation to bulk matter is transistors, light emitting diodes, superfluidity (whereby some liquids lose their viscosity at very low temperatures), solid-state lasers, and semiconductor lasers.

A famous line of attack on reductionism by the holists, when talking about a large collection of particles, is the adage: "Whole is more than the sum of its parts." At the level of the Schrödinger equation itself, this is incorrect as the discussion above clearly demonstrated: the Schrödinger equation of a whole is exactly the sum of its parts. At the level of the *solution* of the Schrödinger equation (and therefore the actual behavior of objects), this must be expected ... *even at the "parts" level*. An electron by itself is a negatively charged particle whose Schrödinger equation does not yield quantized energy levels; a proton by itself is a positively charged particle whose Schrödinger equation also does not yield quantized energy levels. Sum up the Schrödinger equations of the parts and you'll get a hydrogen atom with quantized energy levels.[7]

So, while it may be true that the whole is more than the sum of its parts, without a careful examination of the parts one cannot explain the "more" of the whole.

Extraordinary Matter

When he discovered the atomic nucleus, Rutherford stumbled upon the most well-kept secret of the universe. With almost all the mass of the atom concentrated in it, the nucleus was thought to hold some weighty attribute demanding further exploration. To dig deeper into the nucleus of the atom required accelerating probes moving much faster than alpha particles whose speed was less than one percent the speed of light.

The first particle accelerator was invented by Ernest Lawrence, an American physicist. The *cyclotron*, in which protons were periodically accelerated by an alternating electric field while circulated by a constant magnetic field, was built in 1931, had a diameter of 2.5 inches, and accelerated protons to about 1%

the speed of light. By 1945, the largest cyclotron had a diameter of 15.3 feet and could accelerate protons to more than half the speed of light. Bevatron, designed to convert the energy of the accelerating proton into mass (via $E = mc^2$) and create an anti-proton, was built in 1954, had a diameter of 135 feet, and could accelerate protons to 98.85% the speed of light.

The latest accelerator, the *Large Hadron Collider*, built by the European Organization for Nuclear Research (CERN) between 1998 and 2008 in collaboration with over 10,000 scientists and hundreds of universities and laboratories, has a diameter of more than 8 km and lies in a tunnel 175 m deep beneath the France–Switzerland border near Geneva. At the present time, it can accelerate a proton to 99.9999991% the speed of light. In the future, this percentage may grow even larger.

From a theoretical standpoint, the very small and the very fast demanded a unification of quantum physics and special relativity. The most obvious unification yielded negative probabilities for some properties of particles— a result that at the time was deemed nonsensical. To avoid this negative probability, Paul Dirac discovered an ingenious mathematical technique and came up with an eponymous relativistic quantum equation in 1928.

The Ψ-function appearing in the Schrödinger equation, which describes the motion of a non-relativistic electron (an electron that moves much slower than light), becomes a collection of four functions in the Dirac equation. Two of those functions are associated with the electron.[8] Some rigorous mathematical reasoning showed that the other two components had to describe a positively charged particle having exactly the same mass as the electron. In 1932, Carl Anderson, unaware of Dirac's prediction, discovered such a particle. This positive electron—or *positron*, as it has come to be known—has the interesting properties that by merely putting it next to an electron, the two completely annihilate each other into what appeared at first to be pure energy.[9] Because of this process, positron—being opposed to the very existence of the electron— has been identified as the *anti-matter* of the electron, or anti-electron.

After the identification of the positron as the antiparticle of the electron, physicists started hunting for other antiparticles. With the development of accelerators, they sought a machine that could produce an *anti-proton*. Simple relativistic calculation revealed that the machine had to accelerate a proton to about 99% the speed of light and impinge it on a second stationary proton. Bevatron, developed at the University of California at Berkley in 1954 and put in operation in 1955, created the antiproton in the same year. It is now known that every elementary particle in existence has an antiparticle with identical mass.

The outright denial of scientific progress is a trademark of New Age gurus who cannot face the disproof and destruction of their belief system in light of concrete evidence. Ignoring the successful unification of special relativity and quantum physics, its explanation of spin, prediction and discovery of positron, prediction and production of anti-proton, Capra says:

> What we need, therefore, for a full understanding of the nuclear world is a theory which incorporates both quantum physics and relativity theory. Such a theory has not yet been found, and therefore we have as yet been unable to formulate a complete theory of the nucleus.[10]

There are only two possible explanations for this false statement. Either Capra doesn't know about the Dirac unification of quantum physics and relativity, or he is deliberately misinforming his readers. Since he has a background in high energy physics, he must have had a course in relativistic quantum mechanics, as any graduate student in that field would have. Therefore, the first explanation does not appear to apply.

The Language of Nature

Galileo is well known for initiating and incorporating the "scientific method" in the investigation of nature. By applying this method to the experimental study of motion, he discovered the law of inertia. What is less known is that, in discovering this law, he had to idealize his experimental setup in the same way that geometers had idealized the notion of a point centuries earlier. The link between physics and mathematics was strengthened in his mind as he derived the formula for the free fall of objects and showed that the path of the trajectory of a projectile is a parabola, one of the conic sections studied in geometry. Galileo was so convinced of the role of mathematics in studying nature that he later wrote:

> [The Book of Nature] is written in mathematical language, and its characters are triangles, circles and other geometric figures, without which it is impossible to humanly understand a word; without these, one is wandering in a dark labyrinth.[11]

The "dark labyrinth," to which Galileo refers in this quote, is the methodology of science prevailing at his time, namely the Aristotelian philosophization of natural enquiry. Aristotle wrote the first book on physics, and although the discipline takes its name from that book, his *Physics* and the current discipline have nothing in common. Reading *Physics*, one indeed finds oneself in a labyrinth of long-winded philosophical arguments that are either so general as to be useless, or, when specific, they are wrong.

As he was sitting under an apple tree looking at the rising Moon in late afternoon, Newton discovered the law of gravity—so the legend goes. What is not a legend is that by calculating the acceleration of the Moon on its path around the Earth and comparing it with the acceleration of an apple as it falls to the ground, Newton was able to unravel the mathematical law of the force of gravity. With the second law of motion already at his disposal, he could now write the first *differential equation* in history as applied to the motion of a planet around the Sun and solve it to prove that Kepler's three laws of planetary motion—which themselves were mathematical formulas involving speeds of the planets and their distance from the Sun—were consequences of the mathematical laws of motion applied to the force of gravity. Galileo's vision seemed to take hold ever so strongly in science: mathematics is indeed the language of Nature.

The connection between physics and mathematics proved to be much more deeply rooted than originally perceived. As physics created new mathematics, such as the theory of differential equations (a branch of mathematics which did not exist before Newton), mathematicians found new toys to play with. Once created, the new mathematics became the property of the human mind, which could now apply logic, intuition, and abstraction to evolve it further. It was in this spirit that Joseph Fourier, the great French mathematician of early nineteenth century, said, "The profound study of nature is the most fruitful source of mathematical discoveries."

What is bewildering is that Nature speaks fluently in even the most abstract mathematical dialects developed solely in the confines of the human brain. One example deserves attention.

Group theory is a branch of mathematics that was invented by a French teenager named Évariste Galois in 1830 to address certain purely mathematical puzzles.* Forty years later, Sophus Lie, a Norwegian mathematician, combined Galois' group theory with calculus—itself born out of the study of Nature—and invented yet another mathematical discipline now called *Lie group* theory. At the time of its invention, no one could have thought that Lie group theory would one day creep into physics. Yet it did! And it did it at the most fundamental level of physical theories: Eugene Wigner recognized certain

---

*See Appendix (page 170) for the interesting history of the development of group theory.

transformations of special relativity as elements of a Lie group. He then applied the Lie group theory to relativity and quantum mechanics and showed that a quantum particle is very generally described by two numbers, the values of its mass and its spin either of which can be zero.[12] Yes, you heard Wigner right! A real particle can have zero mass.

This kind of interplay between physics and mathematics has bewildered scientists and philosophers alike and prompted Einstein to say, "The most incomprehensible thing about nature is that it is comprehensible." He had in mind, of course, a comprehension of nature in terms of mathematics, as Galileo had pointed out centuries earlier. Einstein made this clear in another one of his famous quotes concerning the question of how "mathematics, being after all a product of human thought, is so admirably appropriate to the objects of reality." Eugene Wigner described the bewilderment profoundly as "the unreasonable effectiveness of mathematics in the natural sciences."

New Physics Befriends New Math

The invention of the accelerators in the 1930s and their rapid development after WWII opened up a new vista for fundamental physics research. The few particles that made themselves known in the early days of nuclear physics were only the tip of an iceberg. By the mid-1950s the accelerators were producing so many new particles that physicists felt like eighteenth century zoologists facing a seemingly endless variety of "animals" to study. And just like the zoologists, they embarked on their classification.

The initial key to this classification was the strength of the particles' interaction with matter. By injecting them into varying thicknesses of lead, physicists could identify two major categories of particles: those that were stopped after moving a short distance, indicating their strong interaction with the lead atoms, were called *hadrons*, from the Greek word *adros* meaning strong. The remaining particles, which happened to be lighter than hadrons, were named *leptons*, derived from the Greek word *leptos* meaning light. It turned out that the animals that crowded the zoo of particle physics were all hadrons, and these were the particles that needed classification to ease their investigation. Leptons were few and far between.

The early study of hadrons made it obvious that they consisted of two major categories based on the value of their spin—which for all quantum particles can be a positive integer $(0, 1, 2, 3, \ldots)$ or a positive half integer $(\frac{1}{2}, \frac{3}{2}, \frac{5}{2}, \ldots)$.[13] The first category of hadrons are *baryons*, which have half-integer spins; the second category are *mesons*, which have integer spins. Baryons seemed to hold a special place among hadrons.

It appeared that in any collision process, as violent as it might have been and as many different kinds of particles it might have created out of the energy of the initial colliding particles, the total number of baryons at the end was equal to that at the beginning. There was a *baryonic charge conservation* at work, very similar to the electric charge conservation: If you subtract the total number of negative baryons from the total number of positive baryons before the collision and do the same after the collision, the two are always equal. In other words, every time a new positive baryon is created in a collision, it should accompany a negative baryon. In general, positive baryonic charge of +1 is assigned to particles and −1 to the corresponding antiparticle. Thus, protons and neutrons have positive baryonic charge of +1; their antiparticles, negative charge of −1. Mesons, on the other hand, could be created and annihilated individually, and they were produced profusely. They didn't carry any baryonic charge, or their baryonic charge was zero.

To go beyond the twofold classification, two physicists, M. Gell-Mann and Y. Ne'eman, used the abstract mathematics of Lie groups to put all the known hadrons of the time into various families, or *multiplets*. There is a crucial difference between Wigner's use and Gell-Mann/Ne'eman's use of Lie groups. In Wigner's case there already was a Lie group, the transformation group of special relativity. Gell-Mann and Ne'eman had to *invent*—or rather pick from an infinite collection of Lie groups—a Lie group for their classification. Gell-Mann called this classification scheme the *eightfold way*. Following the mathematicians' practice, Gell-Mann and Ne'eman constructed an abstract geometric plane having a horizontal and a vertical axis, the latter being the S-axis, or the *strangeness* axis, because some hadrons behaved contrary to what was expected of them—they had a strange behavior! So, particle physicists had to invent yet another charge: the *strangeness charge*.

Some collection of points of this plane—which turned out to be vertices and certain points on the sides of regularly shaped figures like equilateral triangles and hexagons—represented some special elements of the Lie group, which the eightfold way identified as particles in a family. The *mathematics* of the eightfold way dictated how many and in what order the particles were to assemble themselves at those vertices. And in a remarkable collaboration of Nature and mathematics, Nature obeyed that diktat. Proton, neutron, and many other hadrons respected the will of the mathematics and arranged themselves on the vertices and centers of hexagons.* And if some points were

---

*See Appendix (page 172) for more detail.

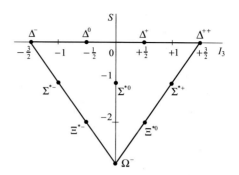

Fig. 9.1 The mathematics of the eightfold way predicted Ω^-

empty, the mathematics, in a Mendeleevesque way, told physicists to look for yet-to-be-discovered particles corresponding to those points.

Among the many particles known in 1961 were four baryons (Δs in Fig. 9.1) which did not seem to belong to a hexagon but could fit in a triangle accepting 10 particles. In the same year, three new particles (Σ^*s in the figure) were discovered, which could also fit in the triangle. A year later at a conference of particle physicists held in Geneva, experimenters reported two new particles (Ξ^*s in the figure) which again fitted nicely in the triangle.

Here is how Ne'eman describes the excitement felt in that conference:

> The creators of the eightfold way, who attended the conference, ... saw the [inverted] pyramid being completed before their very eyes. Only the apex was missing .... Before the conclusion of the conference Gell-Mann went up to the blackboard and spelled out the anticipated characteristics of the missing particle, which he called 'omega minus' (because of its negative charge and because omega is the last letter of the Greek alphabet).[14]

After hunting for it for a couple of months, the omega minus (Ω^-) was captured in one of about 100,000 photographs taken by a group of 33 physicists at the Brookhaven National Laboratory at the end of 1963. In February 1964, the group officially announced the discovery of the Ω^-.

It is worthwhile to pause and think about this story. Forced by a highly abstract mathematical idea, a theorist predicts a particle that *should* exist in nature. He gives all the necessary properties of the particle so that experimenters learn how to look for it ... They do, and they eventually find it. How could this be? How can mathematics, especially the mathematics of Lie groups, which is after all the product of the human mind, find its way into the workings of nature. It is a mystery that Einstein pondered about and Wigner called "the unreasonable efficacy of mathematics." It is a mystery that we may

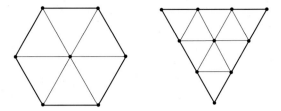

Fig. 9.2 All multiplets of the eightfold way are made up of a fundamental shape that happens to be a triangle

never solve. But one thing appears to emerge from the history of physics: the language of nature is mathematics, and the deeper we dig into the unknown, the more abstract the mathematics becomes.

Quarks

Whenever a phenomenological theory[15] such as the eightfold way succeeds, physicists instinctively look for the deeper, more fundamental, and invariably more reductionist reason behind its success. Dalton's search for the deeper, more fundamental understanding of why chemical substances always mix in exact proportions to yield a compound led him to the reductionist concept of atoms. Planck's search for the deeper, more fundamental reason behind his phenomenological equation for the black body radiation led him to the epitome of reductionism: electromagnetic quanta.[16]

In the case of the eightfold way, the clue to this instinctive drive and a more fundamental understanding of particles comes from the Lie group theory itself. It turns out that all the multiplets used in the eightfold way can be built from a single shape: triangle, as shown in Fig. 9.2 for the hexagon and triangle. Murray Gell-Mann and George Zweig, working independently at Caltech in 1964, came up with the idea that the triangular mathematical building block of the eightfold way ought to be reflected in Nature as three distinct fundamental particles, from which all the hadrons are built. Gel-Mann, borrowing from James Joyce's *Finnegans Wake*, called these particles *quarks*. On the basis of this quark model, all baryons are made up of any combination of three quarks: the *up* quark, the *down* quark, and the *strange* quark, while mesons consist of a quark and an antiquark.*

---

*The Appendix has a section on the quark model (on page 174). It is not hard to read, and it is fun to figure out how to build hadrons from quarks.

I hope I have motivated you to appreciate the unimaginable intimacy between mathematics and Nature. Think about it! A branch of mathematics that didn't even exist in the mid-nineteenth century, finds use in the successful classification of particles produced in man-made accelerators, to the point that a physicist, based on that very mathematics, predicts Ω^-, eventually to be found in an accelerator laboratory. As if that were not incredible enough, the mathematics of Lie groups tells theoretical physicists that all those hadrons are just composites of three fundamental particles: quarks. As weird as these quarks may seem, Nature told physicist, they *are* real particles. And Nature was right.

The quark model of hadrons appeared to have some issues initially. Unlike atoms and nuclei, which would let go of their constituents when persuaded with sufficient amount of energy of an impinging particle, proton stuck to its quarks regardless of the amount of energy carried by the attacking particle. What appeared to be coming out of a proton was hadrons. The increase in energy resulted only in an increase in the number and/or the energy of hadrons. The hadrons-begetting-hadrons idea was the basis of bootstrap hypothesis. The real explanation of *quark confinement* had to wait until certain theoretical ideas, collectively called *gauge theory*, came into being in the late 1960s and early 1970s.

The Third Stage?

To the disappointment of my readers, I have to insist on the inseparability of physics and mathematics; and I hope the preceding narrative eases that disappointment. I can soothe your desire to learn physics by saying "you don't really need math to understand physics"—as do many physics enthusiasts and, unfortunately, professional physicists—but I'll be dishonest with you. And I'd rather tell you the truth, as sour and discouraging as it may be, than make you feel good … as Fritjof Capra does not hesitate to do.

Since mathematics is alien to Eastern theosophy, Capra connives to make it irrelevant to physics. He divides scientific research into three stages. The first two stages consist of gathering experimental evidence (observation) and correlating that evidence with mathematical symbols (theory). Every scientist agrees on these two stages. But now Capra adds another stage:

> … eventually, [physicists] will want to talk about their results to nonphysicists and will therefore have to express them in plain language. This means they will have to formulate a model in ordinary language which interprets their mathe-

matical scheme. ... the formulation of such a verbal model, which constitutes the third stage of research, will be a criterion of the understanding they have reached.[17]

This quote, despite its innocent appearance, is a powerful artifact crafted to legitimize Capra's argument for the parallel between Eastern mysticism and modern physics. The "third stage" of the scientific process, which he introduces here and uses repeatedly in his book, is entirely self-fabricated. The obligation forced on physicists to convey their science to the public is meant to equate the inaccessibility to the public of the mathematical formulation of physical theories with the imposed unspeakability of the Eastern mysticism.

The translation of modern physics into "plain language" is an impossible task because of the sophistication of the mathematical framework and experimental gadgetry employed in current physics research. Needless to say, at earlier times, the required level of sophistication was minimal. The first authoritative work on electricity and magnetism was William Gilbert's book *De Magnete*. Published in 1600, this book contained the results of the frontier research in the field, yet it was all written in an ordinary language, Latin. Anybody educated in Latin and interested in the subject could read the book and understand most of it. Two hundred seventy-three years later, James Clerk Maxwell published *A Treatise on Electricity and Magnetism*, which nobody except a handful of physicists and mathematicians could understand. And the degree of sophistication and specialization has increased exponentially ever since. Nowadays, irrespective of how educated one is, or how much interest one may have in physics, one cannot understand the results of theoretical and experimental discoveries unless one knows the language of physics, namely mathematics and the experimental techniques.

If you want to truly understand Persian poetry, you'll have to learn Farsi. Persian poets are not held responsible for translating their poetry into English or any other language. Similarly, physicists should not be held responsible for translating their work into ordinary languages. While it was possible for Edward FitzGerald to write a great (but not perfect) English translation of Rubaiyat of Omar Khayyam, it is impossible to translate nature's language into any human language.

So, what are physicists to do? Should they simply ignore the laymen, and put the blame on them for not learning the language? Quite the opposite! Due to its inaccessibility, it is the duty of physicists to make as much of their knowledge available to the public as possible. This demands an expertise in the art of translation, for which no physicist is trained and not every physicist is trainable. Fortunately, every generation of physicists has a handful of such "artists" who convey the excitement of discoveries to the public by writing books and articles in ordinary languages and, nowadays, by developing multimedia presentations. The implicit purpose of crudely translating the sophisticated theoretical and experimental techniques in ordinary languages is recruitment. For the discipline to survive, the younger generation ought to be informed of its beauty, excitement, applications, and relevance, in the hope that a fraction of that generation will "carry the torch." However, this is not, by any stretch of imagination, a stage of physical research.

What about the communication of the results to fellow physicists? Isn't that done in an ordinary language, and therefore, a stage—albeit an implicit one—of the development of the discipline? Communication to fellow-physicists, at its most fundamental root, is an evolutionary instinct which tells a physicist that his/her discovery should be made available to the scientific community for further progress. Even the Egyptian and Babylonian priests/scientists knew that understanding the universe was not the task of a single person or a single generation; that knowledge gained by one person should be disseminated to the rest of the community and recorded for posterity so they can start at the new level of understanding and advance that level.

Two physicists, whose professional tongue is the common language of mathematics, can talk to each other even though they may have different mother tongues. Mathematics, being a language, is expressed in symbols. Fortunately, the symbols used by physicists have been more or less universalized, so that almost all physicists adhere to the same set of symbols to express their mathematics. An American physicist, writing in this, and only this, language can convey his/her thoughts to a Chinese physicist. However, if the equations were interspersed with a minimum amount of ordinary language, it would make a much smoother and faster reading. That is why physics articles do contain some ordinary language—which has turned out to be English. But the structure of the English used in physics articles is so simple and the vocabulary so meager that almost all the physicists in the world master it very quickly.

The bottom line is that communication among physicists takes place in the language of mathematics; ordinary language plays the role of a catalyst in this communication; and physicists are not obligated to translate their discovery to the public. Capra's "third stage" is only a Machiavellian ruse fabricated for the sole purpose of drawing a parallel between Eastern mysticism and modern physics.

Leptons

The zoo of hadrons prompted physicists to employ highly abstract mathematics to first classify them, and then discover the fundamental particles—up, down, and strange quarks—of which they were composed. The world of leptons is nothing like the overcrowded world of hadrons. Since leptons do not participate in the strong nuclear interaction, they cannot form particles as do quarks. So, there are only a handful of leptons, and despite subjecting them to violent collisions, no structure has been found inside leptons. As far as we know, all leptons are fundamental particles, just like quarks.

The first lepton, the electron, was discovered in 1897 as one of the constituents of atoms. It is negatively charged, has spin $\frac{1}{2}$, and like proton, is absolutely stable, in the sense that it never disintegrates (meaning that once produced, it lives forever). The second lepton, the *muon*, was discovered in 1947. Except for its mass, which is about 200 times that of the electron, there is no other appreciable difference between the two. Muon is so much like an electron that it is sometimes called a "fat electron." Its enormous mass is a good source of energy for its disintegration. In fact, unlike electron, which cannot find a lighter charged particle into which to decay, the muon can easily decay into an electron plus other particles. What are these other particles?

An analysis of muon decay reveals that there are two other electrically neutral leptons accompanying the electron. These leptons escaped detection for a long time, but eventually they were identified as *neutrinos*.[18] The two neutrinos produced in a muon decay interact differently with matter. When the energy of one of them is transformed into matter upon impact with other particles, a positron is produced, while the other neutrino always produces a muon. Because of this distinction, particle physicists call the first one an electron anti-neutrino, and the second one a muon neutrino.

That's it! Only four leptons, two electrically charged and two neutral. At least that was the story until the 1970s when a "cultural revolution" shook the physics of fundamental interactions.

10

The Standard Model

In the classification of particles in the last chapter, one abundant particle, photon—the particle of light—was conspicuously missing. Photon is neither a hadron nor a lepton; it belongs to a third class of particles called *gauge bosons*, the byproducts of *gauge theories*. According to quantum field theory, fundamental particles exert forces on one another by exchanging gauge bosons. The process of repulsion of two electrons, for example, is described as the exchange of photons between the two electrons.

A clear understanding of gauge theories requires some knowledge of *symmetry*, a concept closely associated with Lie groups. Intuitively, we know that a hexagon is more symmetric than a square. The rigorous reason is that a hexagon can be rotated about its center by a multiple of 60° and still retain its original shape. There are altogether six such multiples. For a square there are only four multiples—of 90°. A circle is the most symmetric two-dimensional shape, because one can rotate a circle by any angle and still get the original figure. This notion of symmetry has an exact mathematical analogue within the framework of Lie group theory: instead of shapes one has mathematical entities (called *fields*) describing particles, and instead of rotations, certain mathematical operations on those fields.

Before continuing, I'll have to make some remarks on "field" to distinguish it from the mystical, non-local, omnipresent field of the New Age gurus, which is somehow associated with soul, spirit, energy, and consciousness. The field of fundamental physics is a mathematical entity describing a *real* fundamental particle, which is associated with the probability of some measurable *property* of that particle. This is very similar to the Ψ-function of the Schrödinger equation, which as we saw in Chap. 4, was associated with the probability of some *property* of a quantum particle. You cannot measure Ψ, only the particle *property* whose probability it describes. Unlike the universal field of modern gurus, every fundamental particle has its own unique field in a given theoretical framework.

Symmetry and Gauge Theory

Our current understanding of the fundamental forces between elementary particles started in 1954 when Chen-Ning Yang and Robert Mills, two theoretical physicists, attempted to generalize the electromagnetic force in the purely mathematical framework of Lie groups. Crucial in this development was their identification of two kinds of symmetry operations: global and local. Think of a field as a sheet of clear transparency. A global symmetry is analogous to applying the same color evenly to all points of the transparency. A local symmetry is putting a blob of color here and there on the sheet. Yang and Mills started with a globally symmetric theory (a mathematical expression involving fields) and found that if they were to make their theory locally symmetric, they had to add new mathematical fields, called *gauge fields*, to their original theory.

The transparency analogy may help understand the origin of the *gauge fields* in local symmetry. Put some red blobs on a sheet of transparency. This is a local symmetry operation, and if your theory is globally symmetric, the local operation does not work in the global context. But suppose you bring a new transparency, and color it red at all places where the first transparency is clear and leave it intact where the first transparency is colored. Now, if you put the two transparencies on top of each other, you get a combination that is evenly colored everywhere. You have restored global symmetry by introducing a compensating transparency, whose coloring was *gauged* by the coloring of the original transparency: wherever there was color in the original, it was left blank in the compensating transparency and vice versa. In Fig. 10.1, the top row shows the local operation on four transparencies, while the bottom shows the compensating—gauge fields of the—transparencies for the top ones.

It turns out that the number of gauge fields required for the localization of a globally symmetric theory increases with the symmetry of the theory: the more symmetric a theory is, the more gauge fields are needed to make it local.

What is the significance of these gauge fields? To answer this question, theoretical physicists lowered the symmetry of the Yang-Mills theory and

10 The Standard Model 143

Fig. 10.1 The "field" affected by some local symmetry operations (top) and their compensating "fields" (bottom)

discovered that the less symmetric theory produced only one gauge field; and this single gauge field could be identified as photon. Since photon was known to mediate the electromagnetic interaction, it was hoped that the gauge fields of the Yang-Mills theory could be made to mediate the weak and strong interactions. However, there were serious theoretical obstacles that had to be overcome before gauge fields could be identified as the mediators of the other forces.

The mathematics of a locally symmetric theory does not allow its gauge fields to have mass. On the other hand, massless mediators give rise only to long-ranged forces like electricity, whose mediator is the massless photon, and gravity, whose mediator is the massless graviton.[1] Weak and strong nuclear forces are known to be short ranged; therefore, their mediators had better be either massive,* or possess some other property that confines their interaction to short ranges. A good part of the late 1960s and early 1970s saw the solution to the dilemma in the context of two ideas: *spontaneous symmetry breaking* (SSB) and *confinement*.†

---

*Here and elsewhere in the book, the word "massive" is used to describe a particle whose mass is not zero. The mass doesn't have to be huge as the word suggests when used in ordinary language.
†All the preceding discussion of gauge theories, as well as what follows, is elaborated in the Appendix (starting on page 175).

Spontaneous Symmetry Breaking

An important concept in any field theory is the vacuum. Lacking any (real) particles, it is the state of the lowest possible energy, and as such, also the most stable of all other states of the field theory.[2] A good analogy to the concept of a vacuum in quantum field theory– in connection with the concept under consideration here—is an inflating balloon.

Consider a perfectly spherical balloon, which we inflate continuously. Assume that as we add more and more air in the balloon, the sphericity of the balloon does not change: every point on and inside the balloon is similar to other points. One says that the equilibrium state (the vacuum) of the balloon and its content are *rotationally invariant.* The inflation cannot, of course, go on forever; the balloon eventually pops. Just before it pops, any part of the balloon is as likely to crack as any other part. However, the process of popping picks out a particular part of the balloon through which the air gushes out. Once this happens (and it happens *spontaneously*), the symmetry of the balloon and the process is broken, and the equilibrium state is no longer rotationally invariant. There is an analogous mathematical process, by which the choice of a vacuum, out of an infinite equivalent possibilities, spontaneously breaks the symmetry of a theory. This process of *spontaneous symmetry breaking* (SSB) has significant consequences in gauge theories.

Jeffrey Goldstone, a British theoretical physicist, showed in the early 1960s that when the *global* symmetry of a theory is spontaneously broken due to the unavoidable choice of a vacuum, some mathematical terms appear in the equations of the theory, which correspond to massless particles, now called the *Goldstone bosons*. Soon after, a number of physicists, including Peter Higgs, another British physicist, applied the idea of SSB to *locally* symmetric theories, and discovered a remarkable property, now called the *Higgs mechanism*. Based on this mechanism, when a theory that is locally symmetric—and therefore, necessarily contains massless gauge fields—is spontaneously broken, some Goldstone bosons disappear, and a corresponding number of the originally massless gauge fields acquire mass. Physicists jokingly say that some of the gauge fields "eat up" the Goldstone bosons and gain weight. Thus, spontaneous symmetry breaking "fattens" the gauge fields and makes them good candidates for short-ranged interactions.

Electroweak Nuclear Force

Yang-Mills' idea of gauge fields was appealing; and Higgs' mechanism of giving mass to the gauge fields added to the appeal. Since the original Yang-Mills theory did not work in practice, physicists ventured into the enchanting land of Lie groups in search of *the* Lie group that held the secret of nuclear forces. The search was initially aimed at unraveling the weak nuclear force. As mentioned in the Appendix, this was tantamount to the correct grouping of particles together. Among the frenzy that ensued Goldstone and Higgs embellishment of Yang-Mills gauge theory, the work of three prominent physicists stood the test of time: Steven Weinberg, Abdus Salam, and Sheldon Glashow.

Weinberg, Salam, and Glashow grouped the quarks and leptons just the way Nature had intended to group them: they put the electron and its neutrino in one group (a *doublet*) and the muon and its neutrino in another. This grouping resulted in three Goldstone bosons and four gauge bosons. Then, by invoking SSB, three of the four gauge bosons "ate up" the three Goldstone bosons and acquired mass. These three massive gauge bosons—two electrically charged and one neutral—were given the symbols W^+, W^-, and Z^0. The fourth gauge boson remained massless and was identified as photon. Weinberg, Salam, and Glashow—in yet another incredible collaboration of Nature and mathematics—came up with not only a satisfactory theory of weak nuclear force, but an unexpected unification of that force with the electromagnetic force, giving rise to the *electroweak nuclear force*.

By the mid-1970s, there were many experimental results which were consistent with this theory. Some of these results could be used to predict the masses of W^+, W^-, and Z^0. By late 1970s, there were so many experimental agreements with the theory that the Nobel Prize was awarded to Weinberg, Salam, and Glashow in 1979. However, no direct evidence of the three massive gauge particles existed at that time.

On January 21, 1983, in a packed auditorium at CERN, the group working on the detection of the $W^{\pm}$ bosons announced their success. They had found a signal exactly at the predicted mass value. A few months later, the Z^0 boson with the expected mass was also discovered at CERN. Today, thousands of sightings of these particles have been reported, and many of their physical characteristics, including their masses, have been measured to many significant figures.[3]

The "Charm" of Mathematics

In the early 1960s there were only two lepton doublets: the electron doublet and the muon doublet. The idea of arranging the quarks into doublets seemed very attractive, and the beta decay* added a strong support to the idea. However, there were only three quarks, and putting the up and down quarks in a doublet—as the beta decay strongly suggested—would leave the strange quark without a partner. Could it be that there was a fourth quark yet to be discovered? That the strange quark indeed had a partner, but the hadron in which it was hiding had not yet been detected? This idea was very *charm*ing mathematically.

November of 1974 was an exciting month for high energy physics. In that month, a new particle was discovered by two groups: one on the east coast at Brookhaven National Laboratory (BNL) led by Samuel Ting who called it J; the other on the west coast at Stanford Linear Accelerator Center (SLAC) led by Burton Richter who called it ψ. This J/ψ particle was identified as a bound state of a charm and an anti-charm and was given a third name: *charmonium*. Once charmonium was discovered and the existence of the charmed quark established, physicists ventured to construct hadrons consisting of a charm and some other quarks, a hadron with a "naked charm."[4] By 1976 the SLAC group was able to produce mesons consisting of a charm and one of the older antiquarks. The 1976 Nobel Prize went to Burton Richter and Samuel Ting for their discovery of the charmed quark.

No sooner had the charmed quark peeped through one detector at SLAC than a new *lepton* started to crawl out of another. In 1974, a group led by Martin Perl was seeing certain unique events from the collision of electrons with positrons at very high energies, which prompted him to suggest that they came from the production, and subsequent disintegration, of *heavy leptons*—leptons that are much heavier than muon. It took Martin Perl and his collaborators four more years to convince the physics community that what they had discovered in 1974 was the τ lepton or *tauon*,[5] and by the end of 1978 all properties of tauon were nailed down. Tauon has its own neutrino and the two form a tauon doublet. Because of its huge mass, tauon can easily decay into a muon and its antineutrino or an electron and *its* antineutrino. Such decay modes have been seen numerously ever since the discovery of tauon. One half of the 1995 Nobel Prize went to Martin Perl for his discovery of tauon. The

---

*See page 177.

other half went to Frederick Reines for another significant work: his detection of neutrinos.

With the word of the discovery of the third lepton doublet getting around, the prospect of yet another quark loomed over many accelerator labs. If charm raised the number of quark doublets to the level of the-then existing two lepton pairs, the discovery of tauon (and its neutrino) intimated a new quark doublet.

As early as 1974, there were mixed signals coming from detectors at Fermilab that hinted at a new hadron. However, these mixed signals were not untangled until 1977, when the discovery of Υ—pronounced *upsilon*—at Fermilab was announced. Υ, ten times heavier than proton, was shown to be a bound state of a new quark and its antiquark. Because of its negative charge, the quark was named *bottom*.*

If there is a bottom quark, there ought to be a *top quark* as well. The discovery of the bottom quark at Fermilab started the "race to the top" in all laboratories capable of accelerating particles and their antiparticles to a sufficiently high speed. With the bottom quark mass at about five proton mass—almost half the mass of the Υ consisting of the equally massive bottom and anti-bottom quarks—physicists anticipated a mass of 10, 15, or 20 proton masses for the top quark. However, cranking up the Υ-production energy by a factor of 4 or 5 yielded no results. Even cranking up the energies to 50 or 75 proton masses showed no sign of the top quark. After 18 years of experimenting, Fermilab reported the detection of a top-antitop bound state at about 360 proton mass in 1995, giving a mass to the top quark that is approximately 180 times proton mass and a 35-fold increase over the mass of the bottom quark.

Higgs Boson

The electroweak theory contains twelve matter fields (six leptons and six quarks) and four gauge bosons that mediate the electroweak force between them. All of these fields (particles) have been confirmed experimentally. And while in the late 1970s a quick succession of discoveries anticipated tauon and its neutrino and the bottom and top quarks, no other matter field has been expected or discovered since then. In fact, there are some strong theoretical arguments and cosmological observations that point to the nonexistence of

---

*The multiplets that—for typesetting purposes only—I have put in horizontal rows, such as (ν_e, e^-) and (ν_μ, μ^-) in the Appendix, are actually put in vertical columns, with the negatively charged member at the *bottom*.

additional quarks or leptons. It therefore appears that we are done. After all, what else is there except material objects and forces among them? Well, there *is* something else.

Remember the *Higgs mechanism*, whereby some of the gauge fields "eat up" the Goldstone bosons and become massive? In order to give mass to the gauge bosons, the Higgs mechanism introduces yet another field (particle) into the theory. This particle is called the *Higgs boson*; sometimes dubbed the "God particle," because its presence in the theory gives mass to other particles. This is the last piece of the electroweak theory left to be discovered. It is predicted to have zero spin and have an extremely large mass, perhaps of the order of 100–150 GeV.

With all the predictions of the electroweak theory having been substantiated, there is little doubt in the minds of the elementary particle physicists that the Higgs boson will eventually be captured. It is only a question of when and where. The laboratory that can pump sufficient amount of energy into its accelerator will be the first to see the "God particle." At this point, all eyes are on Geneva, where the physicists at the European Center for Particle Physics are analyzing the data obtained in the successful runs of the largest accelerator ever, the *Large Hadron Collider* (LHC).

There are very strong indications that the Higgs boson has already been seen at about 130 proton masses. However, bona fide experimental physicists are extremely cautious in announcing their results.[6] Although they have reached a confidence level at which they can claim that the chance of their being wrong is one in a few thousand, they are waiting to boost their confidence to a level where their being wrong has a probability of one in a million. Thus, we are on the verge of the discovery of the last piece of the puzzle of the electroweak theory as physicists at the European Center for Particle Physics analyze data of the experiments run in 2011, the centennial of Rutherford's discovery of the atomic nucleus.

I wrote the preceding two paragraphs in my lecture notes for a course I was teaching in the early years of 2000, months before July 4, 2012, when the group working on the production of the Higgs boson at CERN announced the discovery of a "Higgs-like" particle with a mass of about 126 GeV in Geneva. I have left those paragraphs in to convey the remarkable predictive power of some physical theories and the faith we physicists have in those theories. I use the word "faith" to describe our conviction in a theory that has been tested in a multitude of harsh experiments and as a deterrence to those who want to propose a new "theory" every time the old theory appears to be inconvenient.

Physics theories are not of the "just-a-theory" variety. They are deeply rooted in the ground of many precise experimental outcomes and exact mathematical statements. Physicists' "faith" in a theory has absolutely no resemblance to a religious faith in God or universal consciousness or spirit, and it differs from the latter in two ways: it has experimental and observational evidence to back it up, *and* when strong evidence indicates the incorrectness of the theory, the physicists abandon their "faith."

Confinement and Strong Nuclear Force

The weak nuclear force describes mostly the disintegration of nuclei. Its short range is due to the mass of the gauge particles intermediating it. What about the strong nuclear force? What kind of force holds the protons and neutrons inside a nucleus? More fundamentally, what holds the quarks inside protons, neutrons and other hadrons? An analysis of the quarks inside certain hadrons led Gell-Mann to introduce the notion of a *color charge* for quarks.[7] It is the force associated with this color charge that holds together the quarks inside a hadron, just as the electric force of the positive nucleus and the negative electrons holds an atom together. However, there are substantial differences between those two forces.

The gauge theory of electroweak interaction put particles in doublets and let a Lie group act on those doublets. The gauge theory of strong nuclear force puts the three color charges of each quark in a triplet and lets a specific larger Lie group act on those triplets. This theory has eight gauge bosons, which intermediate the force between the quarks. The gauge bosons of the strong nuclear force are themselves colored; and this color makes them attract each other and the quarks into a very "sticky glue," thus the name *gluons* for these gauge bosons. The theory of quarks interacting via gluons is called *quantum chromodynamics* or QCD. The outcome of the quark-gluon interaction is *confinement*.

In contrast to the electrical force between two electrically charged particles, which decreases—thus making the separation easier—as you pull them apart, the color force between two quarks *increases* when you pull them apart: quarks strongly resist separation. This is also true of the gluons themselves. So, while the quarks and gluons materially reside inside hadrons, one cannot forcefully knock them out, because any attempt at separating them will encounter a strong resistance which only increases with separation. That is why isolated quarks and gluons cannot be found: isolation requires very large separation, and large separation means exponentially increasing resistance.

The energy aimed at separating the quarks will instead create other quarks and antiquarks—due to the relativistic equivalence of mass and energy, $E = mc^2$—which pair up among themselves and the original quarks to form new hadrons; thus, the plethora of particles created in the 1950s and 60s, when physicists tried to probe inside protons.

QCD, discovered by David Gross, Frank Wilczek, and David Politzer in 1973, has been so successful in explaining and predicting the outcomes of hadronic collisions and structures that the 2004 Nobel Prize in physics was awarded to its discoverers. When the electroweak theory and quantum chromodynamics are put together, the result is called the *standard model* of the fundamental particles and interactions.[8]

The Taoist's Denial

The experimental and theoretical developments in nuclear and subnuclear physics in the last century as outlined in this and the last chapters are arguably the greatest achievements of mankind. Not only have we gained a fairly thorough, mathematical, and verifiable understanding of the microscopic world, but, because of this gain, we can now decipher the workings of no less a mystery than the creation of the universe itself. The comprehension of the big bang, a colossal accelerator created 13.8 billion years ago when the universe was a microscopic object, requires a thorough knowledge of the fundamental particles and forces. All this progress has been, to the New Age "theorists," nauseously reductionistic. They would rather hide their heads in the sand of defunct holistic theories like bootstrap than accept the reality of the power of reductionism in understanding the world around us.

The first edition of *The Tao of Physics* came out in 1975. By that time, elementary particle physicists had discovered a promising venue to pursue their interest: gauge theories. Bootstrap and S-matrix had left physics and entered the land of mysticism. The fifth edition came out in 2009. In the intervening 34 years, a lot of progress was made in the world of *real* physics: prediction and discovery of charm, production of electroweak gauge bosons, discovery of tauon and its neutrino, prediction and discovery of bottom and top quarks, and countless other confirmations of the standard model. Yet there is absolutely no mention of any of these tremendous successes in the fifth edition.

Capra ignores QCD's successful explanation of quark confinement and questions the existence of quarks because they have not been seen in isolation. He laments the fact that "most physicists still hang on to the idea of basic building blocks of matter which is so deeply ingrained in our Western scientific tradition," and takes pride in stating that he has "made it clear that [he considers] the bootstrap philosophy as the culmination of current scientific thinking ... and as the one that comes closest to Eastern thought. ... [and wonders why] not a single Nobel Prize has so far been awarded to any of the outstanding physicists who contributed to the S-matrix theory."[9] No statement has made it clearer what New Ager gurus think of modern physics than this quote: if it doesn't pass the Eastern test, it is not physics, even though all relevant experiments and observations attest to its validity.

Surprisingly, Capra seems to accept the *idea* of quarks, and wants to incorporate them within the S-matrix theory and the bootstrap hypothesis. However, the quarks that he has in mind are non-physical (spiritual maybe?). He gives the reader the impression that a major breakthrough has occurred in bootstrap hypothesis *recently* which can "account for the observed quark structure without any need to postulate the existence of physical quarks."[10] The references that he gives for this breakthrough are two articles that *he himself* wrote; one in 1979 in a pedagogical—as opposed to research—journal of physics, and another in 1981 in an obscure journal called *Re-Vision*, which according to its website "publishes articles of theory, research, and teaching practices related to basic writing." Perhaps Capra is relying on the nonchalance of readers to citations and references. But claiming a breakthrough and inculcating in the mind of his readers the implication that his work deserves a Nobel Prize is crossing a line.

Progress in high energy physics is not the only thing that New Age gurus refuse to lift their head from under the sand to see. Capra seems to deny any success attributed to modern mainstream physics. On page 63 of the fifth edition, he says that while the special theory of relativity has been confirmed experimentally, "the general theory [of relativity, GTR] has not yet been confirmed conclusively." It is said that the proof is in the pudding. Every time Capra uses his GPS, he is proving the general theory. Global Positioning System is indispensably based on the general—as well as the special—theory of relativity. If the manufacturers of GPS do not take into account the impact of relativity on time and distance measurements, their product will be useless. This is only one of the many confirmations of the GTR. Black holes, supernovae, expansion of the universe, the big bang, and gravitational waves are merely a handful of the verified predictions—and therefore, conclusive confirmation—of the GTR.

Holistic philosophy teaches the interconnectedness of everything and is opposed to any notion of basic building blocks of matter. Physicists, nevertheless, pursue the methodology that began with Galileo, at the heart of which is reductionism. I close this chapter by listing the Nobel Prizes in physics awarded—since the year in which Capra saw "cascades of energy coming down from outer space, and the atoms of the elements and those of [his] body participating in [a] cosmic dance of energy"—to nearly sixty physicists who "still hang on to the idea of basic building blocks of matter."

1969   Murray Gell-Mann "for his contributions and discoveries concerning the classification of elementary particles and their interactions."

1972   John Bardeen, Leon Neil Cooper and John Robert Schrieffer "for their jointly developed theory of superconductivity, usually called the BCS-theory."

1976   Burton Richter and Samuel Chao Chung Ting "for their pioneering work in the discovery of a heavy elementary particle of a new kind."

1979   Sheldon Lee Glashow, Abdus Salam, and Steven Weinberg "for their contributions to the theory of the unified weak and electromagnetic interaction between elementary particles, including, inter alia, the prediction of the weak neutral current."

1980   James Watson Cronin and Val Logsdon Fitch "for the discovery of violations of fundamental symmetry principles in the decay of neutral K-mesons."

1984   Carlo Rubbia and Simon van der Meer "for their decisive contributions to the large project, which led to the discovery of the field particles W and Z, communicators of weak interaction."

1985   Klaus von Klitzing "for the discovery of the quantized Hall effect."

1988   Leon M. Lederman, Melvin Schwartz, and Jack Steinberger "for the neutrino beam method and the demonstration of the doublet structure of the leptons through the discovery of the muon neutrino."

1989   Norman F. Ramsey "for the invention of the separated oscillatory fields method and its use in the hydrogen maser and other atomic clocks."

1990   Jerome I. Friedman, Henry W. Kendall, and Richard E. Taylor "for their pioneering investigations concerning deep inelastic scattering of electrons on protons and bound neutrons, which have been of essential importance for the development of the quark model in particle physics."

1992 Georges Charpak "for his invention and development of particle detectors, in particular the multiwire proportional chamber."

1995 "for pioneering experimental contributions to lepton physics."
Martin L. Perl "for the discovery of the tau lepton."
Frederick Reines "for the detection of the neutrino."

1999 Gerardus 't Hooft and Martinus J.G. Veltman "for elucidating the quantum structure of electroweak interactions in physics."

2001 Eric A. Cornell, Wolfgang Ketterle and Carl E. Wieman "for the achievement of Bose-Einstein condensation in dilute gases of alkali atoms, and for early fundamental studies of the properties of the condensates."

2004 David J. Gross, H. David Politzer, and Frank Wilczek "for the discovery of asymptotic freedom in the theory of the strong interaction."

2008 Yoichiro Nambu "for the discovery of the mechanism of spontaneous broken symmetry in subatomic physics;"
Makoto Kobayashi and Toshihide Maskawa "for the discovery of the origin of the broken symmetry which predicts the existence of at least three families of quarks in nature."

2012 Serge Haroche and David J. Wineland "for ground-breaking experimental methods that enable measuring and manipulation of individual quantum systems."

2013 Francois Englert and Peter W. Higgs "for the theoretical discovery of a mechanism that contributes to our understanding of the origin of mass of subatomic particles, and which recently was confirmed through the discovery of the predicted fundamental particle, by the ATLAS and CMS experiments at CERN's Large Hadron Collider."

2014 Isamu Akasaki, Hiroshi Amano and Shuji Nakamura "for the invention of efficient blue light-emitting diodes which has enabled bright and energy-saving white light sources."

2015 Takaaki Kajita and Arthur B. McDonald "for the discovery of neutrino oscillations, which shows that neutrinos have mass."

2016 David J. Thouless, F. Duncan M. Haldane and J. Michael Kosterlitz "for theoretical discoveries of topological phase transitions and topological phases of matter."

2017 Rainer Weiss, Barry C. Barish and Kip S. Thorne "for decisive contributions to the LIGO detector and the observation of gravitational waves."

2019 James Peebles "for theoretical discoveries in physical cosmology;" Michel Mayor and Didier Queloz "for the discovery of an exoplanet orbiting a solar-type star."

2020 Roger Penrose "for the discovery that black hole formation is a robust prediction of the general theory of relativity;"
Reinhard Genzel and Andrea Ghez "for the discovery of a supermassive compact object at the centre of our galaxy."

2022 Alain Aspect, John F. Clauser and Anton Zeilinger "for experiments with entangled photons, establishing the violation of Bell inequalities and pioneering quantum information science."

Pitifully, despite such remarkable achievements of *real* modern physics, the public is manifoldly exposed to its distorted version as told by New Age gurus. Even educated citizenry has been taught to delight in the infestation of modern physics with Eastern mysticism because it finds a fictitious scientific basis for its fascination with yoga, meditation, and acupuncture. And when reputable physicists, in the footsteps of the founders of quantum physics, encourage this infestation, we are bound to nail the coffins of critical thinking and, eventually, reality, as we are witnessing now.

11

Epilogue

A painting hung on a wall inconspicuously for years has recently been identified as possibly the work of Michelangelo. The tempera, the layering, and the pigments all seem to point to the style of the Renaissance artist. An unknown poem, discovered centuries after it was written, gives away the name of its famous writer to a scholar with expertise in the style of that poet. Given a printed copy—to eliminate identification from the composer's handwriting—of a newly discovered piece of music, a musicologist specializing in Beethoven can not only identify it as the work of the master, but also determine the period of the composer's creativity in which it belongs. The creation of an artist bears the style of its creator. Not so in science!

Einstein presented his general relativity (GR) field equation on 25 November 1915 to Prussian Academy. Five days earlier, David Hilbert, the great German mathematician, had presented a talk containing the same equation to the Royal Academy of Sciences in Göttingen. A historian of physics, given the printed version of the two equations, could not tell which one is Einstein's and which one Hilbert's. The GR field equation does not bear the identity of its creator.

GR field equation is not an isolated incidence of simultaneous independent discoveries made by different physicists. Although only Einstein's name is attached to special relativity, other physicists contributed to its discovery substantially. The mathematical relation that connects the physical quantities measured by two observers moving relative to each other is called the *Lorentz transformation* after Hendrik Lorentz, the Dutch physicist who discovered the relation independently of Einstein. Henri Poincaré, the great French mathematician, discovered a more general version of Lorentz transformation independent of Lorentz and Einstein. Heisenberg and Schrödinger discovered two different versions of the same quantum physics independently. Quarks were proposed by Gell-Mann and Zweig. Charm quark was discovered simultaneously at Brookhaven National Laboratory in Upton, NY and Stanford Linear Accelerator Center in Stanford, CA.

Scientists study material quantifiable objects of nature. The information with which they are equipped to study nature is identical: it is the accumulation of scientific knowledge available to them. The material world from which they extract information in their areas of research is also identical. The combination of materiality, identical knowledge, and identical objects of inquiry tends to culminate in identical scientific discoveries by different people. Oftentimes Nobel Prizes are shared by two or three scientists for making the same discovery. This is unique to sciences. Examples from other branches than physics include the independent formulation of calculus by Isaac Newton and Gottfried Wilhelm Leibniz; discovery of oxygen by Carl Wilhelm Scheele, Joseph Priestley, and Antoine Lavoisier; advancement of the theory of the evolution of species by Charles Darwin and Alfred Russel Wallace.[1] In no other field of human endeavor do we see a creation made independently by two creators. The notion of two Mona Lisas painted independently by two artists, or two Hamlets written by two independent writers, or two Eroicas composed by two different musicians is manifestly absurd.

Science Is Detached from Scientist

There are two sides of a scientist that are mutually exclusive: their science and their personal character. A scientist can be Christian, Muslim, Jew, Buddhist, or atheist; they may be Austrian, Belgian, Yemeni, or Zambian; they can have political views that are liberal, conservative, socialist, communist, or be indifferent to politics; they may like classical, jazz, hip hop, or reggae music; they could have lived thousands of years ago or flourished recently. Despite all the variation in the personality of scientists, their science is unique and bears no sign of their character—or their *style*. The principle of buoyancy discovered by a Greek scientist more than two thousand years ago is as valid today as it was then. The laws of electromagnetism discovered by an evangelical Scottish physicist hold as true now as in 1865 when he published them. The theory of gravity proposed by a German Jew in 1915 is still as valid as then.

The validity ascribed to their science cannot be attributed to the social, political, religious, or philosophical viewpoints of scientists. Newton's religious beliefs were so unorthodox that they bordered on occultism. Lavoisier's political views were on the wrong side of history causing his execution by guillotine at the age of fifty. Einstein's letter to President Roosevelt, warning that Germany might develop atomic bombs and suggesting that the United States should start its own nuclear program, prompted action by Roosevelt, which eventually resulted in the Manhattan Project. Einstein later regretted signing the letter, telling *Newsweek* magazine that "had I known that the Germans would not succeed in developing an atomic bomb, I would not have lifted a finger." The majority of scientists who worked on the Manhattan Project regretted helping with the project after they learned of the bombing of Hiroshima and Nagasaki.

Scientific ideas are detached not only from the worldview of their discoverers, but even from the discoverers' mistaken interpretation of those ideas. Schrödinger believed that the mathematical function in his equation described some sort of matter waves, but Max Born correctly identified it as probability amplitude. Being firmly against introducing probability in fundamental physics, Schrödinger is said to have remarked that he might not have written his paper had he known of the consequences. As thousands of experiments and inventions have shown, he was wrong and Born was right: Schrödinger's equation is completely decoupled from Schrödinger.

An indicator of the separation of science from scientist is the difference in the venues of the publication of their science versus their unscientific ideas and opinions. The former, typically a journal, goes through a peer review process[2] whose rigor and reliability defines the caliber of the journal. The review process of the latter, typically a trade book, is considerably more lax. Bohr published his pioneering work in *Philosophical Magazine*; his mystical viewpoints were published in *Atomic Physics and Human Knowledge*, a trade book. Heisenberg published his uncertainty principle in *Zeitschrift für Physik*; his philosophical viewpoints were published in *Physics and Philosophy* and a few other trade books. Schrödinger published his groundbreaking equation in *Annalen der Physik*; his mystical viewpoints were compiled in *What is Life?* and *Mind and Matter*, yet another trade book.

The detachment of science from the scientists' worldviews is also plainly displayed in the textbooks from which professional scientists learn their subjects. Consider consciousness, of which the founders of quantum physics talked profusely and Eugene Wigner promoted its role to the foundation of understanding quantum physics when he said that it was not possible "to formulate the laws of quantum mechanics in a fully consistent way without reference to the consciousness."[3] If consciousness is so fundamental in the formulation of quantum mechanics, then any textbook from which professional physicists learn the subject must, at some point, introduce consciousness and incorporate it in the discussion of quantum physics. Yet, no quantum mechanics textbook talks about consciousness. There is no chapter or even a section in which the author tells the reader, "here is where consciousness enters into the understanding of quantum physics;" or "to understand such and such technical concept, you, the graduate student earning a PhD in physics, must have a full comprehension of consciousness;" or that to invent laser, LED, and microchip, the inventors had to impose consciousness at some stage of their invention.

The public, being more familiar with artistic creation, is prone to confusing the truth of scientific ideas with the worldviews of their discoverers. And pop-spiritualists take full advantage of this confusion. Just as Beethoven's style cannot be separated from his music, they posit, John Wheeler's mystical views cannot be separated from his physics, and his observer-created reality is on as firm a ground as his black hole.[4] In case you don't remember the masterful application of this trickery by Chopra, I urge you to go back to page 27 and reread his introduction to the latest version of *Quantum Healing*.

Don't "Make Sense"!

One of the young participants in the discussion of religion and science in the Solvay Conference of 1927 and recounted by Heisenberg[5] was Paul Dirac, a brilliant mathematical physicist who proved the equivalence of Schrödinger's quantum mechanics and Heisenberg's so-called *matrix mechanics*. Dirac's monumental contribution to physics—the one that laid the groundwork for the current standard model of the fundamental forces and particles—was the unification of special relativity and quantum physics, in the process of which he created a new branch of mathematics called *spin geometry*.[6] He is arguably in the top five greatest physicists of the twentieth century, surpassed only by Einstein and possibly Planck.

Dirac's work not only clarified the mysterious concept of spin but *predicted* the existence of antimatter. The matter-antimatter annihilation into what *appears* to be "pure energy"[7] is used opportunistically by modern gurus to prove "scientifically" the unity and sameness of matter and soul, spirit, healing energy, Qi, …. Why is it then that these mystics, whose writings and speeches are filled with mystical quotes associated with Heisenberg, Schrödinger, Pauli, and Bohr, never mention Dirac? His remarks in that gathering may shed some light on the answer. Heisenberg recollects the conversation:

"I don't know why we are talking about religion," [Dirac] objected. "If we are honest – and scientists have to be – we must admit that religion is a jumble of false assertions, with no basis in reality. The very idea of God is a product of human imagination. It is quite understandable why primitive people, who were so much more exposed to the overpowering forces of nature than we are today, should have personified these forces in fear and trembling. But nowadays, when we understand so many natural processes, we have no need for such solutions. … the postulate of an Almighty God … leads to such unproductive questions as why God allows so much misery and injustice, the exploitation of the poor by the rich and all other horrors He might have prevented. … Religion is a kind of opium that allows a nation to lull itself into wishful dreams and so forget the injustices that are being perpetuated against the people."[8]

Modern gurus are masters of cherry-picking quotations from famous physicists and linking the quotations to the science of those physicists. They never mention the opposition of physicists like Einstein and Planck to mysticism and their warning against the direction in which the popular writings of scientists such as Bohr, Heisenberg, Schrödinger, and Pauli were going. The matter-antimatter annihilation takes place in the context of relativistic quantum physics, which was invented by Dirac. Why don't modern gurus associate the transformation of matter into energy—no, into real material particles that *carry* energy—to Dirac, who did not believe in any kind of spirituality, including the God of the West or the cosmic consciousness of the East?

The weirdness of quantum physics, arising from superposition and the relation of the wave function to probability,[9] which initially derailed the comfort that physicists had come to feel with the determinacy of classical physics, has become the norm in physics. This normalcy is not unlike the normalcy reached in classical physics by the end of the eighteenth century. That century began with Newton's new mathematics of differential equation to describe the motion of planets around the Sun. And to "make sense" of the mathematics, he invoked the hand of God to set up the initial conditions of the planets. By the end of that century, physicists had gotten so used to mathematics that Laplace could say, "Sire, I had no need of that hypothesis.", when Napoleon remarked that there was no mention of God in Laplace's *Mécanique Céleste*.

The generations of physicists, who came after the tumultuous years of the early twentieth century, accepted probability as part of the language of nature. They have succeeded in applying quantum theory to the tiniest subatomic particles and the largest galaxies and have discovered fantastic new phenomena and astonishing gadgets along the way. By the end of the twentieth century, physicists realized that they had no need of the hypothesis of consciousness that the founders introduced to "make sense" of quantum mechanics.

Richard Feynman, one of the most creative physicists of the latter part of the twentieth century and hardly mentioned by modern gurus, embodied the opposition to "making sense" through an invocation of consciousness or philosophy. Unlike his predecessors, he detested any philosophical interpretation of quantum physics. In his 1964 MIT Messenger lecture, Feynman talks about the probabilistic nature of the behavior of electrons and the difficulty in imagining what an electron is. He then goes on to say,

> The difficulty really is psychological and exists in the perpetual torment that results from your saying to yourself, "But how can it be like that?" which is a reflection of uncontrolled but utterly vain desire to see it in terms of something familiar. I will not describe it in terms of an analogy with something familiar; I will simply describe it. ... I think I can safely say that nobody understands quantum mechanics. So do not take the lecture too seriously, feeling that you really have to understand in terms of some model what I am going to describe, but just relax and enjoy it. I am going to tell you what nature behaves like. If you will simply admit that maybe she does behave like this, you will find her a delightful, entrancing thing. Do not keep saying to yourself, if you can possibly avoid it, "But how can it be like that?" because you will get 'down the drain', into a blind alley from which nobody has escaped. Nobody knows how it can be like that.

It was precisely the "desire to see it in terms of something familiar," or in Heisenberg's words, the desire to "make sense" of quantum physics, that drove the founders 'down the drain' of Schopenhauer's philosophy and Eastern mysticism. After almost forty years of the old generation of physicists trying to "make sense" of quantum physics, Feynman is telling his listeners about the danger in doing so. And to emphasize Galileo's dictum of mathematics being the only sensible language of Nature, he is said to have told his student who was insisting on engaging him in a philosophical dialogue, "Shut up and calculate!"

Another member of the new generation of quantum theorists, who proved the non-locality of quantum physics and who, like Feynman, is ignored by modern gurus, is John Bell. Like Feynman, he avoids trying to "make sense" of quantum physics, but rather concentrates on its mathematical probabilistic structure. On the observer-created reality, Anthropic Principle, information, and Eastern theosophy, Bell says,

11 Epilogue   161

> ... let me argue against a myth ... that quantum theory had undone somehow the Copernican revolution. From those who made that revolution we learned that the world is more intelligible when we do not imagine ourselves to be at the centre of it. Does not quantum theory again place 'observers' ... us ... at the centre of the picture? ... And from some popular presentations the general public could get the impression that the very existence of the cosmos depends on our being here to observe the observables. But I see no evidence that it is so in the success of contemporary quantum theory.
>
> So, I think it is not right to tell the public that a central role for conscious mind is integrated into modern atomic physics. Or that 'information' is the real stuff of physical theory. It seems to me irresponsible to suggest that technical features of contemporary theory were anticipated by the saints of ancient religions ... by introspection.
>
> The only 'observer' which is essential in orthodox practical quantum theory is the inanimate apparatus which amplifies microscopic events to macroscopic consequences. Of course this apparatus, in the laboratory experiments is chosen by the experimenters. ... But once the apparatus is in place, and functioning untouched, it is a matter of complete indifference ... according to ordinary quantum mechanics ... whether the experimenters stay around to watch or delegate such 'observing' to computers.[10]

If John Bell were alive in 2022, the Nobel Committee would have had to replace one of the three recipients of the prize in that year by John Bell because of the rule that no more than three persons can receive the prize in the same field, and because the three recipients of 2022 Nobel Prize were experimental physicists verifying *Bell's inequality*.

Unfortunately, rational thoughts, such as those of Dirac, Feynman, Bell, and other physicists like them, don't make it to the psyche of the public. The public is fascinated by both the enchantment of mysticism and the convenience of science-based technology. It is, however, repulsed by the esotericism of science itself. There is therefore nothing more appealing to the public than a science that is fundamentally tainted with mysticism. When the founders of quantum physics and other notable scientists like Eddington and Jeans introduced mysticism and consciousness in physics in early twentieth century, the public rejoiced in make-believing that physics was not as abstruse as some physicists portrayed it to be, but "a rich, profound venture which had become inseparable from philosophy."[11] Despite the admission of guilt by some of the founders and the fierce opposition of giants like Einstein and Planck, the belief in the presence of consciousness at the quantum level shaped the way the public came to understand quantum physics, not just in that period, but for all generations ever since.

The unjustified mystical halo around quantum physics is today ceremoniously fortified by the worrisome mushrooming of pseudoscience celebrities who are manifestly ignorant about the essence of science in general, and quantum theory in particular. Because the public arena is deluged with misinformation produced by profit-hungry publishers and media moguls, the feeble sound of reason drowns in that arena. The only way science can be separated from New Age mysticism is behind the relatively protected walls of *classrooms*. And it falls on us science educators to prioritize science literacy in our curricula. Let's teach the next generation what science is, how it progresses, how it is detached from non-material concepts like soul or cosmic consciousness, how its distinct branches tie together like a web, and how different it is from what New Age gurus are publicizing it to be. Because pop-spiritualists' most pernicious weapon against science is mixing scientists' outlook with their science, an indispensable component of what we teach our students ought to be the truth that science has nothing to do with political, social, and philosophical opinions of scientists themselves: *It is the message that counts, not the messenger.*

Appendix

Chapter 1

Homeopathy

Samuel Hahnemann is considered the creator of homeopathy, the belief that the same substance that causes the symptoms of a disease in healthy people will cure similar symptoms in sick people. He believed that if the substance is repeatedly diluted in alcohol or distilled water, followed by forceful striking on an elastic body, the resulting remedy cures the disease.

The dilution is measured in "C"s. Each C represents a dilution of one part in one hundred. (Why 100, no one knows.) For example, a 2C dilution means that the original sample was diluted in a volume of water or alcohol 100 times larger than the sample, and then the resulting volume was diluted in a volume 100 times larger than its size. Suppose that the original sample has a volume of 5cc and contains 10,000 molecules of the substance. Pour this 5cc in a volume of 495 cc to get 1C. Now 10,000 molecules are distributed (evenly) in 500cc, and 5cc of this contains 1/100 of the molecules. So, after 1C, a volume of 5cc would contain $10{,}000 \times 1/100$ or only 100 molecules. By the same reasoning, after 2C, a volume of 5cc would contain only one molecule. And after 3C, only one of the 5cc-samples carries the lonely molecule and the other ninety nine 5cc-samples would be pure water or alcohol.

Continuing this line of reasoning, you can see that if the original 5cc contained a million molecules, after 3C, a 5cc sample would contain only one molecule; if the original 5cc contained a hundred million molecules, after 4C, a 5cc sample would contain only one molecule. In general, an original 5cc

Fig. A.1 This bottle is most likely one of the 99...99 (thirty eight 9s) bottles that have zero molecule of the drug!

sample with 100^n molecules, would yield a sample with one molecule after nC. A typical real sample contains about $10^{22} = 100^{11}$ molecules. Therefore, an 11C sample contains only one molecule; only one of the 12C sample carries that molecule, the other 99 are empty. This means that the chance of picking the container that has the molecule in it is 1/100. Suppose we pick one of the containers and go through another dilution step. Now the chance of picking the molecule is $(1/100)(1/100) = 1/10,000 = 100^{-2}$. In other words, only one of the ten thousand 13C (note that $10,000 = 100^{13-11}$) samples carries a molecule, the other 9999 are empty. Continuing, only one of the million (100^{14-1}) 14C sample carries a molecule, the other 999,999 are empty. A 15C homeopathic drug is not uncommon. So, 99,999,999 out of a hundred million (100^{15-11}) samples sold in homeopathic drugstores are empty, and the 100 millionth carries only one molecule! The bottle that you see in Fig. A.1 has a label of 30C. So, it is most likely one of the $100^{30-11} - 1 = 10^{38} - 1$ or 99...99 (thirty eight 9s) bottles that have zero (yes, zero) molecule of the drug!

Hahnemann hypothesized what he called miasms as the underlying causes of disease and that homeopathic remedies—achieved by dilution—eliminated these. There was absolutely no evidence for such a hypothesis, and the treatment had no more success than a placebo. Nevertheless, mostly due to the harsh practices of the medicine of the time, which included bloodletting and purging, homeopathy became popular in Europe and the US in the nineteenth century. By 1900, there were 22 homeopathic colleges and 15,000 practitioners in the United States.

Nevertheless, from the very beginning, homeopathy was criticized by mainstream science, both in Europe and in the US. Even the leading homeopathists of Europe abandoned the practice of administering infinitesimal doses and no longer defended it. The last school in the U.S. exclusively teaching homeopathy closed in 1920. However, the rise of the New Age movement in the 1960s and 1970s saw a surge in the popularity of the Far Eastern culture, and with it, an exponential increase in the practice of the alternative medicine, including homeopathy.[1]

Chapter 4

Coin Probabilities

Let's first enumerate the eight possible outcomes of tossing three coins: HHH, HHT, HTH, THH, HTT, THT, TTH, TTT. So, there is only one possibility of getting three heads, three possibilities of getting two heads, three possibilities of getting one head, and one possibility of getting zero head. When you toss n coins, you have two possibilities for the first coin, two possibilities for the second coin, two possibilities for the third coin, and so on. Thus, the number of possible outcomes is $2 \cdot 2 \cdot 2 \cdots 2$, that is 2 times itself n times, or 2^n. This agrees with the known cases of one coin ($n = 1$ and $2^1 = 2$ possibilities), two coins ($n = 2$ and $2^2 = 4$ possibilities), and three coins ($n = 3$ and $2^3 = 8$ possibilities).

The next question is, "in how many ways can I get m heads when I toss n coins?" Let $C(n, m)$ stand for the number of ways. Then, it can be shown that

$$C(n, m) = \frac{n!}{m!(n-m)!},$$

where the symbol ! (called *factorial*) after an integer means "multiply all integers from 1 up to and including that integer." For example, $1! = 1$, $2! = 1 \cdot 2 = 2$, $3! = 1 \cdot 2 \cdot 3 = 6$, $4! = 1 \cdot 2 \cdot 3 \cdot 4 = 24$, and so on. Although it may not make sense to you, it turns out that $0! = 1$—it will make sense once you see some examples.

Let us verify the formula above for the known cases of 1 and 3 coins (you can do the case of two coins yourself if you want some practice with the formula). For one coin ($n = 1$), we can have one head ($m = 1$) or no head ($m = 0$), and

$$C(1, 1) = \frac{1!}{1!(1-1)!} = \frac{1!}{1!0!} = \frac{1}{1 \cdot 1} = \frac{1}{1} = 1,$$

$$C(1, 0) = \frac{1!}{0!(1-0)!} = \frac{1!}{0!1!} = \frac{1}{1 \cdot 1} = \frac{1}{1} = 1.$$

So, as expected, there is one way that we can have either one or no H in tossing one coin. For three coins ($n = 3$), we can have three heads ($m = 3$), two heads ($m = 2$), one head ($m = 1$), or no head ($m = 0$), and

$$C(3, 3) = \frac{3!}{3!(3-3)!} = \frac{3!}{3!0!} = \frac{6}{6 \cdot 1} = \frac{6}{6} = 1,$$

$$C(3, 2) = \frac{3!}{2!(3-2)!} = \frac{3!}{2!1!} = \frac{6}{2 \cdot 1} = \frac{6}{2} = 3,$$

$$C(3,1) = \frac{3!}{1!(3-1)!} = \frac{3!}{1!2!} = \frac{6}{1\cdot 2} = \frac{6}{2} = 3,$$

$$C(3,0) = \frac{3!}{0!(3-0)!} = \frac{3!}{0!3!} = \frac{6}{1\cdot 6} = \frac{6}{6} = 1.$$

So, again as expected, there is one way that we can have three or no H's and three ways that we can have one or two H's in tossing three coins. I hope the calculations above have convinced you that setting 0! equal to 1 makes sense.

I have gone through details of calculating the familiar cases to build your trust in the formula, because, as you'll see, the formula will lead to very strange results that are hard to believe and you may want to discard it if it were not for the fact that those strange results come from the same formula that yields the familiar believable results.

The probability $P(m, n)$ of getting m heads in tossing n coins is obtained by dividing $C(m, n)$ by the total number of possible outcomes, 2^n:

$$P(n, m) = \frac{C(m, n)}{2^n} = \frac{n!}{m!(n-m)!2^n}.$$

For example, the probability of getting 6 heads when tossing 10 coins is

$$P(10, 6) = \frac{10!}{6!(10-6)!2^{10}} = \frac{10!}{6!4!2^{10}} = \frac{3{,}628{,}800}{(720)(24)(1024)} = 0.205.$$

Thus, when you toss 10 coins the chance of getting 60% of them to turn up H is 20.5%.

In probability we talk about the number of successes in an experiment. If you perform an experiment N times and the probability that you get what you are looking for (probability of success) is p, then the number of successes is

$$\text{number of successes} = Np.$$

For example, if your experiment is tossing 10 coins and success occurs when 6 heads show up, then in repeating the experiment 10,000 times,

$$\text{number of successes} = (10{,}000)(0.205) = 2050,$$

i.e., (on average) you'll get 6 heads 2050 times.

Sometimes we describe a probability as "the odds of being successful is one in …" where the dots are some (usually large) number. The "one" in the quote is the number of successes. So, if you set the number of successes in the formula above equal to one, you get $N = 1/p$ as the number corresponding to the dots in the quote above. For example, if the probability of success is 0.000000015, "the odds of being successful is one in about 67 million" because

$$N = \frac{1}{p} = \frac{1}{0.000000015} = 66{,}666{,}667.$$

Chapter 5

The Double-Slit Experiment

With Ψ_1 and Ψ_2 respectively representing the probability amplitudes for the first and second slits, if only the first slit is open, the total amplitude is just Ψ_1 (because $\Psi_2 = 0$), and the probability is $|\Psi_1|^2$. This gives rise to a single bright image on the photographic plate. The same result holds if only the second slit is open, except that now the probability is $|\Psi_2|^2$; but this probability is identical to $|\Psi_1|^2$ if the two slits are identical. When both slits are open, the probability is $|\Psi_1 + \Psi_2|^2$, which is not just the sum $|\Psi_1|^2 + |\Psi_2|^2$, as the reader recalls from high school algebra: the square of the sum of two quantities is the sum of the squares of the two quantities *plus* twice the product of the two quantities. It is this last term that gives rise to the interference.

Readers familiar with complex numbers and trigonometry can appreciate the following mathematical derivation, which shows precisely where the dark regions come from. It turns out that one can write $\Psi_1 = Ae^{i\phi_1}$ and $\Psi_2 = Ae^{i\phi_2}$, where A is a real number, $i = \sqrt{-1}$, and ϕ_1 and ϕ_2, also reals, are called the *phase angles* of the two amplitudes and depend on the distance from the slits to the point of interest on the photographic plate, and therefore, on the location of that point on the plate. It is not hard to show that $|\Psi_1|^2 = A^2$, $|\Psi_2|^2 = A^2$, and $|\Psi_1 + \Psi_2|^2 = 2A^2(1 + \cos(\phi_1 - \phi_2))$. As you move on the photographic plate, $\phi_1 - \phi_2$ changes. When it is zero or a multiple of 360°, the total probability will be $4A^2$, corresponding to bright bands. When $\phi_1 - \phi_2$ is an odd multiple of 180°, the total probability will be zero, corresponding to dark bands.

Chapter 9

Sizes and Distances of Moon and Sun

Size of Moon Relative to Earth: Part (a) of Fig. A.2 shows the Earth-Moon system as the Moon goes through a total eclipse. The sun is on the left very far away—so that its rays are parallel—shining on the Earth-Moon system and creating shadows to the right of the two objects. The Moon is assumed to revolve in the clockwise direction around the Earth and is about to enter the shadow of the Earth. Part (b) shows a snapshot of the Moon as it enters behind the Earth—as time passes, Moon sinks further and further in the dark region behind the Earth until it completely disappears and finally emerges from the other side (the bottom of the gray circle in part (b)). The shadow of the Earth covering the lower half of the Moon can be compared with the Moon itself, and the size of the Moon relative to the size of the Earth can be estimated. By taking a photograph of the Moon (or drawing it very carefully, as was done by the Greek astronomers) and using some elementary geometry, one can get a very good estimate of the diameter of the Moon relative to the diameter of the Earth.

Far Is Small: You have no doubt observed that when an object moves away from you, it appears smaller. This is obviously not because its actual size decrease, but because the angle subtended by the object at your eyes gets smaller. For an object that is extremely far away, namely that its actual size is much smaller than the distance from your eyes, there is a simple relation

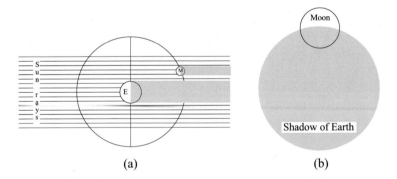

Fig. A.2 (a) The Moon is about to enter the shadow of the Earth. (b) As the Moon is entering the shadow of the Earth, the image of the Earth on the Moon can be compared with the Moon itself, and the size of the Moon relative to the size of the Earth can be estimated

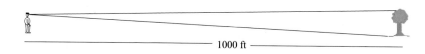

Fig. A.3 From the distance and the angle you can find the height of the tree

that connects the three quantities, angle, size, and distance: the product of distance and angle in *radians*² equals the size. So, if you know any two of these quantities, you can find the third.

The angle drawn from the eyes of the person in Fig. A.3 is 2° or 0.035 rad. Multiply that by 1000 feet to get 35 feet as the height of the tree.

Distance of Moon: Now draw a(n imaginary) line from your eyes to a point on the edge of a full Moon and another line to the point diametrically opposed to the first point and measure the angle between those lines. You'll see that the angle is about 0.5° or 0.0087 rad. Dividing the diameter of the Moon (one-third Earth diameter) by 0.0087 gives the Earth-Moon distance as 38 Earth diameters. Aristarchus' result was lower than this because his estimate of the Moon angle was smaller than 0.5°.

Distance of Sun: There are two half-moons, separated by about two weeks, in a single revolution of the Moon around the Earth (M_1 and M_3 in Fig. A.4). By measuring the time the Moon takes to go from M_1 to M_3 and comparing it with the time from M_3 back to M_1, one can determine the angle α in the figure marked at E. With the help of the figure, you can convince yourself that α is one-fourth the difference between the larger arc of the lunar orbit and the smaller arc.

For us to see a half moon, the center of the Moon should be at such a location on its orbit that the line connecting the center of the Earth to the center of the Moon is perpendicular to the line joining the center of the Sun to the center of the Moon. This makes the line connecting the center of the Sun to the center of the Moon tangent to the lunar orbit. It is clear from the figure that the farther the Sun is from the Earth, the smaller the angle α, and that α is also one of the angles of the right triangle formed by the Earth, Moon, and Sun.

The Earth-Moon distance, Earth-Sun distance, and α are related: if you know two of them, you can find the third one. Aristarchus had already found the Earth-Moon distance. He found α by measuring the Moon's travel times from M_1 to M_3 and from M_3 to M_1. Knowing these two numbers, he could deduce the Earth-Sun distance. Measurement of α turns out to be

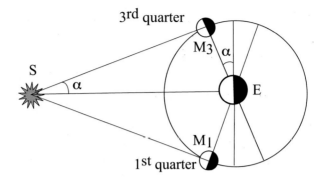

Fig. A.4 From the difference between the arc lengths M_1M_3 and M_3M_1, one determines α, which gives $\overline{SE}$ in terms of $\overline{ME}$

extremely difficult because the difference between M_1-M_3 time and M_3-M_1 time is immeasurably small, which, in turn is due to the fact that the Earth-Sun distance is much larger than the Earth-Moon distance. Nevertheless, Aristarchus' measurements led him to conclude that the Sun is about 20 times farther away from Earth than Moon is.

French Teenager Invents New Math

A polynomial is the sum of different powers of an unknown, usually denoted by x, each power multiplied by a number. The highest power of the polynomial is called its *degree*. When a polynomial is set equal to zero, it becomes a *polynomial equation*. Solving this equation is tantamount to finding the values of x—in terms of the numbers appearing in the polynomial—that satisfy the equation.[3]

The *quadratic equation* is a polynomial equation of degree two. Its solution has a long history going back to the Babylonians. The current form of the solution was found by Muhammad ibn Musa al-Khwarizmi, the Persian mathematician and astronomer of the ninth century to whom the founding of algebra is attributed and whose Latinized name is the origin of the word "algorithm."

The solution to the *cubic equation* had to wait until the sixteenth century when four Italian mathematicians not only solved it, but also introduced complex numbers. As a bonus to their effort, they also solved the *quartic equation*—the polynomial equation of degree four.

For the next three hundred years, mathematicians tried to solve the *quintic equation*, the polynomial equation of degree five, and failed repeatedly. Then in the late 1820s, a French teenager by the name of Évariste Galois proved that, unlike the equations of lower degrees, it is impossible to solve a polynomial equation of degree five and higher. In order to prove his proposition, Galois had to invent—yes, invent—a new branch of mathematics now known as *group theory*.

The theory of equations, starting with al-Khwarizmi and ending with Galois, is a purely mathematical field. Even its origin lies in mathematics itself. When the Babylonians posed the problem of solving the quadratic equation, they were dealing with a purely mathematical curiosity. The problem was neither coming from Nature,[4] nor was it intended to solve a puzzle related to the natural world.

Lie Groups: Group theory is a branch of mathematics born out of mathematics itself. A byproduct of group theory was *representation theory*, in which groups are represented by concrete entities more receptive to numerical manipulation.

Differential equations, on the other hand, was a gift from physics to mathematics. Newton awarded the first differential equation (DE) to mathematicians in late seventeenth century. By mid-nineteenth century, the theory of DEs had turned into one of the most active areas of mathematical research. A drawback of the theory of DEs was that, while mathematicians invented ingenious methods of solving particular DEs, a general systematic study of them was missing. This changed in the 1870s.

In his attempt at steering the study of DEs away from ad hoc solutions conjured up for particular DEs and toward a systematic analysis, Sophus Lie, a Norwegian mathematician, incorporated Galois' group theory with calculus and created yet another branch of mathematics now called *Lie groups*. Lie's creation was so abstract and so purely mathematical that some mathematicians described it as something that would never find an application in physics. ... They were wrong!

On Spin

When it was first discovered, a "spinning electron" was thought to be much like a spinning top. Just as a top can spin with small or large angular speed about an axis that is not necessarily vertical, so can an electron spin with arbitrary speed and about an axis with arbitrary direction. However, it was soon realized that neither the speed nor the direction of the spin of a quantum particle is arbitrary: they both take on only discrete values. This *quantization* of spin is another crowning achievement of the twentieth century physics, whose hallmark is the melding of the most abstract mathematics with the most concrete experimental results.

Relativity theory incorporates a *transformation* that connects physical quantities measured by two observers moving relative to one another. Eugene Wigner recognized this transformation as a Lie group, which could be *represented* as quantum mechanical states of particles. Thus, by combining relativity, Lie group theory, and quantum mechanics in 1939, Wigner found that a particle is a quantum mechanical object that has two numbers associated with it, one defines its mass and the other its spin, *either of which could be zero*.*

Wigner showed that as a multiple of $\hbar$ (Planck constant divided by 2π, also called the *reduced Planck constant.*), the spin s of a particle can be either a nonnegative integer or half integer. Once s is determined, the number of directions[5] in which it can point is $2s+1$ if the particle is massive, and 2 if it is massless. For electron, proton, and neutron, all massive and having a spin of $\frac{1}{2}$, the number of directions is two, usually called *up* and *down*. For photon, having a spin of 1, the number of directions of spin is not 3 ($= 2 \times 1 + 1$), but 2—one along the direction of motion and one opposite to it—because photon is massless.

The spin s is always a multiple of $\frac{1}{2}$. When the spin of a particle is an odd multiple ($\frac{1}{2}, \frac{3}{2}, \frac{5}{2}, \ldots$), the particle is called a *Fermion*, and when its spin is an even multiple (0, 1, 2, 3, ...) the particle is called a *Boson*. Fermions obey Pauli's Exclusion Principle: No two identical Fermions can be in the same quantum state. This explains the periodic table of elements.

The Eightfold Way

Nature had already provided a Lie group for Eugene Wigner to work with: the relativistic transformations relating two observers. The success of Wigner's work encouraged other physicists to apply Lie groups to other areas of fundamental physics. One such area that seemed promising was the classification of hadrons. The difference was that here, the Lie group was not given. It had to be chosen from among an infinitude of Lie groups. Gell-Mann and Ne'eman,

---

*Yes, there is such a thing as massless particle! Its existence is as solid as the mathematics that proves it.

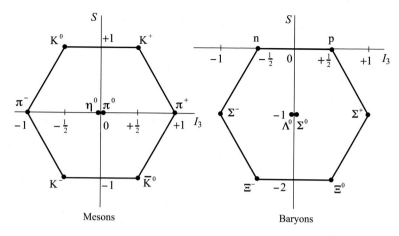

Fig. A.5 The meson octet (left) and baryon octet (right) consist of eight particles each

found the correct Lie group to put all the known hadrons of the time into various families, or *multiplets*, dictated by the mathematics of that particular Lie group.

The geometrical representation of that Lie group on a plane consisted of regular polygons, most notably a hexagon and a triangle. The hexagon consisted of eight special points, which in the scheme invented by Gell-Mann and Ne'eman, represented eight hadrons, thus named *octet*. The triangle had ten special points and accepted ten hadrons, thus named *decuplet*.

The two oldest known baryons, proton (p) and neutron (n) fell into a *baryon octet*, a collection of eight baryons assembled on the vertices and center of a hexagon, as shown on the right in Fig. A.5. The seven known mesons at the time also assembled themselves on the seven points of a hexagon (six on the perimeter and one, π^0, at the center) as shown on the left in the figure.

It was remarkable that for many of the existing hadrons, the eightfold way found appropriate positions in its two-dimensional plane. Moreover, since none of the multiplets were completely filled, the eightfold way *predicted* some particles along the same line that Mendeleev's periodic table predicted certain elements a century earlier.[6] For instance, in 1961, when the eightfold way was proposed, only seven mesons were known. The eightfold way predicted the existence of an eighth meson—now known as η_0 and shown at the center of the hexagon on the left in the figure—along with many of its physical properties. It also predicted certain characteristics of the baryon octet members—shown on the right of the figure—which were not known at the time.

Table A.1 Properties of quarks. The antiquarks have charges opposite to the above

Quark	Electric charge	Baryonic charge	Strangeness charge
u	$+\frac{2}{3}$	$+\frac{1}{3}$	0
d	$-\frac{1}{3}$	$+\frac{1}{3}$	0
s	$-\frac{1}{3}$	$+\frac{1}{3}$	-1

The Quark Model

Their name is not the only thing that is weird about quarks. They have unusual electric and baryonic charges as well. Every known particle has an electric charge that is an *integral* multiple of the charge of a proton e, which in terms of the ordinary unit of charge C, called *Coulomb* (named after the French physicist, Charles-Augustin de Coulomb, who discovered the mathematical formula for the force between two electric charges), has the very small value of $e = 1.60217663 \times 10^{-19}$ C. However, quarks have *fractional* charges. The charge of the up quark is $+\frac{2}{3}e$. The down and strange quarks each has a charge of $-\frac{1}{3}e$. Similarly, the baryonic charges of all known baryons are either $+1$ or -1, but for quarks it is $\frac{1}{3}$, and for antiquarks $-\frac{1}{3}$. The important properties of quarks are summarized in Table A.1. The charges of antiquarks have opposite signs.

The hadrons can be explained by the following simple rules: *All baryons are constructed out of three quarks (antibaryons out of three antiquarks). All mesons are made up of a quark and an antiquark. The spin of each quark (or antiquark) is $\frac{1}{2}$*. Let's use this rule and Table A.1 to find the quark content of the hadrons on the two octets on page 173 and the decuplet on page 135.

The baryons in the hexagon on the right have three quarks. Proton (p) and neutron (n) are on the horizontal axis which crosses the strangeness axis at the origin. So, they have no s quark. The only combination of u and d that can give a zero charge is udd. That's the quark content of n. For p it is uud. The three Σs have strangeness -1. So, they all have one s quark, which has an electric charge of $-\frac{1}{3}$. Therefore, the content of Σ^- is sdd, of Σ^0 is sdu, and of Σ^+ is suu. The two Ξs have strangeness -2. So, they both have two s quarks. Therefore, the content of Ξ^- is ssd and of Ξ^0 is ssu.

All the mesons in the hexagon on the left have one quark and one antiquark. K^+ and K^0 cross the strangeness axis at $+1$. So, they have one anti-s quark, which has an electric charge of $+\frac{1}{3}$. Therefore, the quark content of K^+ is

$\bar{s}u$ and that of K^0 is $\bar{s}d$. The three πs have strangeness 0. So, they have no s quark. Therefore, the content of π^- is $\bar{u}d$, of π^0 is $\bar{u}u$ or $\bar{d}d$, and of π^+ is $\bar{d}u$. This quark content indicates that π^- is the antiparticle of π^+ (or vice versa) and π^0 is the antiparticle of itself (or has no antiparticle). The two Ks have strangeness -1. So, they both have one s quark. Therefore, the content of K^- is $s\bar{u}$ and of $\bar{K}^0$ is $s\bar{d}$. This quark content indicates that K^- is the antiparticle of K^+ (or vice versa) and $\bar{K}^0$ is the antiparticle of K^0. Unlike π^0, the quark content of $\bar{K}^0$ is not the same as K^0. So, $\bar{K}^0$ and K^0 are two distinct particles. The particle η^0, next to π^0, has strangeness zero, not because it has no s, but because it has one s and one $\bar{s}$.

I can continue with the quark content of the baryons on the decuplet. But now that you have learned the tricks of the trade, I believe you'll have fun doing it yourself. Use the table of quarks and keep two things in mind: (1) the electric charge of particles are given as superscripts $^+$, $^-$, and ^{0}, and (2) the vertical axis gives strangeness.

Chapter 10

On Gauge Theory

In a general gauge field theory, the fields associated with particles are grouped together in a multiplet dictated by the Lie group chosen for the theory. Furthermore, any member of a group can turn into any other member while emitting a gauge boson (to be eventually exchanged with another particle with which that member is interacting).

Picture a point—the vertex—at which the initial member of the multiplet, represented by a line, meets two other lines representing the final member of the multiplet and a gauge field. As an example, suppose a Lie group requires three particles in each *triplet*: (p_1, p_2, p_3). In Fig. A.6, six of the nine possible vertices are shown. The symbol G stands for gauge field and the subscripts differentiate between the nine possible gauge fields. In the first diagram, p_1 sends a gauge boson, G_{11}, toward a particle with which it is interacting without losing its identity: it remain p_1 in the end. In the second diagram, p_1 turns into p_2 and sends a gauge boson, G_{12}, toward the particle with which it is interacting.

Now suppose there are three other particles that, in the same Lie group, can be grouped in a triplet, say (q_1, q_2, q_3), as the p-particles do. Then the gauge bosons emitted by the p-particles can be absorbed by the q-particles to constitute a physical process. Figure A.7 shows two processes: scattering and

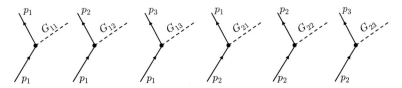

Fig. A.6 For the triplet (p_1, p_2, p_3), there are actually nine vertices. These are six of those nine

Fig. A.7 The diagram on the left is the scattering of p_1 and q_2, producing q_1 and p_2. The diagram on the right describes the decay (or disintegration) of p_1 into p_2, q_1, and $\bar{q}_2$

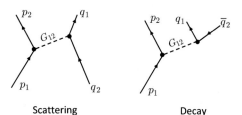

decay. The figure on the left describes a process in which p_1 impinges on q_2—or q_2 on p_1—and two new particles, p_2 and q_1 are created. The bottom of the figure is the initial state and the top, the final state of a process.

The figure on the right shows a p_1 at the bottom and three particles, p_2, q_1, and $\bar{q}_2$ at the top. This is a decay process in which a single particle (here p_1, assumed to have more mass than the sum of the masses of the particles on top) in the initial state disappears and a number of particles (here three: p_2, q_1, and $\bar{q}_2$) are created in the final state. The bar on the symbol for a particle designates its anti-particle.

In these so-called *Feynman diagrams*, particles have an arrow pointing up and anti-particles have an arrow pointing down. In a Feynman diagrams, you can rotate a line around a vertex: the diagram on the right is obtained by rotating the q_2 line of the left diagram counterclockwise … and turning it into the antiparticle of q_2! So, the diagram on the right represents the decay of a p_1 into a p_2, a q_1, and an anti-q_2

As a specific example of the general gauge theory outlined above, consider the weak nuclear force. The formalism of the Lie group of the Weinberg-Salam-Glashow (WSG) theory puts the electron and its neutrino—the symbol for neutrino is ν—in one *doublet*, (ν_e, e^-), and the muon and its neutrino in another, (ν_μ, μ^-). Therefore, based on the general gauge field theory, an electron or a muon can turn into its neutrino and a gauge particle and vice versa.[7] This grouping produces four gauge particles altogether, two electrically neutral, one positively charged, and one negatively charged. Spontaneous symmetry breaking gives mass to three of these: the two charged ones, which are now denoted by W^+ and W^-, and one of the neutral ones, which is

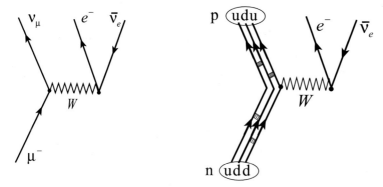

Fig. A.8 The left diagram shows the decay of muon. The diagram on the right is the neutron beta decay

given the symbol Z^0. These give rise to the short-ranged weak nuclear force. The other neutral gauge particle remains massless and is identified as photon, responsible for the long-ranged electromagnetic force. Therefore, WSG theory unifies the weak nuclear force with electromagnetism into the *electroweak* force.

The electroweak theory can readily explain the decay of the muon. The left diagram in Fig. A.8 shows a muon, μ^-, emit a (virtual) unobservable W^- and become a muon neutrino, ν_μ. This is possible—as discussed on page 175—because muon and its neutrino are members of the same doublet. Then the emitted W^- turns into an electron and its (anti)neutrino (also possible because electron and *its* neutrino are in the same doublet). Thus, the overall effect is for the muon (shown at the bottom, corresponding to the initial state) to decay into an electron, a muon neutrino, and an electron anti-neutrino (shown at the top, corresponding to the final state).

The theory can also explain the nuclear beta decay, in which a neutron of a radioactive nucleus turns into a proton, an electron, and an antineutrino. The explanation entails the quark content of a proton (two up quarks and a down quark: *uud*) and a neutron (one up quark and two down quarks: *udd*), and the theoretical assumption that an up quark pairs up with a down quark to form a doublet just as the leptons and their neutrinos do. In the diagram on the right of Fig. A.8, one of the down quarks, d, emits a W^- and becomes an up quark, u, while the emitted W^- turns into an electron and its antineutrino in exactly the same way as in the muon decay. Thus, the overall effect is for the down quark to decay into an electron, an up quark, and an electron anti-neutrino. Once the down quark in the neutron, n, becomes an up quark, the neutron becomes a proton, p.

Notes

Chapter 1: Prologue

1. See http://bit.ly/2X1JbaG, especially minute marks 8:30 to 8:50. The last time I checked, Oprah had blocked the site. However, it is available on Wayback Machine at https://bit.ly/3UQqKpK.
2. Williamson, M. (2019). *Politics of Love.* HarperOne. p. 224, Williamson, M. (1996) *A Return to Love*, HarperOne. p. 66.
3. http://bit.ly/2kdNroS.
4. Yoga would probably be useful as a form of exercise in a healthy lifestyle when done within reasonable limits. Regular physical exercise has shown to be beneficial for multiple conditions and for general health as a whole in multiple validated studies. Yoga isn't anything special or an exception to that. It is the 'cure for all' claims, with no evidential basis, that are disturbing and harmful to the collective mentality of the public, who is led to believe that yoga has some special power because it strengthens not just the body but also the soul. The mental harm becomes manifold when yoga's association with the supernatural is claimed to have been proven by quantum physics. See https://sciencebasedmedicine.org/the-yoga-rct/ for a scientific evaluation of yoga.
5. Hofstadter, R. (1966). *Anti-Intellectualism in American Life.* Vintage Books.
6. https://conspirituality.net/redpilled/.
7. https://www.bbc.com/news/world-55957298 and https://lat.ms/3FCVTCp.
8. http://whatstheharm.net/index.html.

9. https://content.time.com/time/health/article/0,8599,237613,00.html.
10. https://bit.ly/472Hjmq.
11. https://quackwatch.org/.
12. https://sciencebasedmedicine.org/.
13. A recent study (http://bit.ly/2VLHUUd) of more than 30,000 people over several decades has shown that taking supplemental vitamins and minerals offers no discernible benefits in terms of reducing risks of death generally, or death from cardiovascular disease and cancers, specifically. In fact, the study found potential harms. Getting high doses of calcium (1000 mg or more per day) from supplements—but not from foods—was linked to higher cancer mortality. Likewise, people taking vitamin D supplements who didn't have vitamin D deficiencies may have higher risks of all-cause mortality and death from cancers.
14. This feature was theoretically predicted by John Bell in 1964, who died before sharing the 2022 Nobel Prize in Physics which was awarded to three experimental physicists who verified Bell's theoretical work.

Chapter 2: From Myth to Philosophy

1. The distortion of a physical occurrence into an outlandish legend when passed orally from one generation to the next should come as no surprise. In the Telephone Game, also known as Chinese Whisper, one person utters a sentence in the ear of a second person who passes it on to a third person, and so on. When the last person is asked to repeat the sentence, it is unrecognizable from the original. This game has been known for a long time and played in parties for fun. In one example in the UK, the sentence "Send reinforcements, we are going to advance." turned into "Send three and fourpence, we are going to a dance."
2. Clark, E. (1953). *Indian Legends of the Pacific Northwest*. University of California Press. pp. 53–55.
Zdanowicz, C. M., Zielinski, G., & Germani, M. (1999). "Mount Mazama eruption: Calendrical age verified …," *Geology*, 27, 621–624.
3. Pellegrino, C. (1994). *Return to Sodom and Gomorrah*. Random House. p. 133.
4. Kraut, Richard, "Plato", *The Stanford Encyclopedia of Philosophy* (Fall 2017 Edition), Edward N. Zalta (ed.), URL = http://stanford.io/37ZkG5s.
5. Plato; Lee, Desmond, *The Republic*, Penguin Classic (2007), pp. 258–259.
6. There are two exceptions, atomism and natural philosophy, both of which shift the primacy from mind to matter. The latter injects the extra ingredient of observation and experimentation in philosophy. Galileo and Newton are credited with natural philosophy, and thus with creating the scientific discipline to which we owe our current civilization. Ironically, mathematics, which Plato thought to be entirely a product of the mind, had its origin in the

observation of the natural world. Arithmetic and geometry progressed only because humankind experimented with counting and measurement of land eons before Plato. That is why mathematics, although a product of the human mind like philosophy, is so different from the wild speculations of philosophy. Like science, at its roots, mathematics is, if not tamed by the solidity of observation, then at least restrained by the rules of logic and inference.
For atomism, see page 125.
7. Armstrong, K. (1993). *A History of God*, Alfred A. Knopf. p. 4.
8. Ibid. p. 7.
9. https://jothishi.com/buddha-on-astrology/.

Chapter 3: Sins of the Fathers

1. See Chap. 4 for the strange characteristics of probability in general.
2. The mushrooming of acupuncture and Ayurvedic clinics, the proclamation of 21 June as the International Day of Yoga by the United Nations on 11 December 2014, and the popularity of medical practices originating in the mythology and theosophy of the Far East, are the unfortunate affirmation of Schopenhauer's prediction.
3. M. Müller, *The Upanishads*, Part 2, Oxford At the Clarendon Press, (1884), p. 318; online at http://www.sacred-texts.com/hin/sbe15/sbe15117.htm.
4. http://bit.ly/2YMLKOI.
5. Quinn, S. (1995). *Marie Curie: A Life*. Simon and Schuster. pp. 208 and 226.
6. Lachapelle, S. (2011). *Investigating the Supernatural: From Spiritism and Occultism to Psychical Research and Metapsychics in France*, 1853–1931. Johns Hopkins University Press. p. 82.
7. Eddington, A. (1920). *Space Time and Gravitation*. Cambridge University Press. p. 182.
8. In this context, a deleterious misconception, which persistently accompanies quantum-physics-mysticism marriage, has to be clarified. The alleged haphazardness embodied in Schopenhauer's philosophy and Eastern theosophy is the property of the *macroscopic* world, the world of our experience. The probabilistic nature of quantum phenomena is the property of the *microscopic* world, the world of subatomic particles. The behavior of a large aggregate of such particles—an aggregate that belongs to the world of our experience—becomes as predictable by quantum physics as the motion of a planet around the Sun, the epitome of classical determinism. Therefore, in the world of our experience, Schopenhauer's philosophy and Eastern theosophy are as different from quantum physics as they are from deterministic classical physics.
9. Bohr, N. (2010). *Atomic Physics and Human Knowledge*, Dover Publications, p. 20.

10. These principles are discussed in some detail later on page 52.
11. Capra, F. (1989). *Uncommon Wisdom*. Bantam Books. pp. 42–43.
12. Heisenberg, W. (1958). *Physics and Philosophy: The Revolution in Modern Science*. Harper and Row Publishers. p. 202.
13. See page 158.
14. Heisenberg, W. (1972). *Physics and Beyond: Encounters and Conversations*. Harper and Row Publishers. pp. 83–85.
15. According to synchronicity, there is no such thing as a coincidence. Coincidences are meaningful although they may appear acausal. Jung introduced synchronicity to "explain" phenomena that could not be explained by a cause-effect relationship:

> It is impossible, with our present resources, to explain ESP [extrasensory perception] … as a phenomenon of energy. This makes an end of the causal explanation as well, for 'effect' cannot be understood as anything except a phenomenon of energy. Therefore it cannot be a question of cause and effect, but of a falling together in time, a kind of simultaneity. Because of this quality of simultaneity, I have picked on the term 'synchronicity' to designate a hypothetical factor equal in rank to causality as a principle of explanation.*

> In other words, one can use synchronicity to "explain" *any* phenomenon that cannot be explained rationally. ESP is one example. But all paranormal phenomena become good candidates for synchronicity: "Difficult, flawed, prone to misrepresentation, [synchronicity] none the less remains one of the most suggestive attempts yet made to bring the paranormal within the bounds of intelligibility. … Indeed, Jung's writings in this area form an excellent general introduction to the whole field of the paranormal."† That Pauli—who has been compared to Einstein for his brilliance—believed in the pseudoscientific idea of synchronicity, is a clear example of the fact that even science geniuses are not immune from claptrap.

16. Enz, C. P. (2002). *No Time to be Brief: A scientific biography of Wolfgang Pauli*. Oxford University Press. p. 426.
17. Pauli, W. (1994). *Writings on Physics and Philosophy*. Springer. p. 163.
18. Enz, p. 540.
19. Schrödinger, E. (1992, 2013). *What is Life? with Mind and Matter & Autobiographical Sketches*. Cambridge University Press. p. 169.
20. Ibid p. 120.

---

*Jung, C. (2014 [1952]). "Synchronicity: An Acausal Connecting Principle", translated by R. F. C. Hull. pp. 3373–3509 in *Collected Works of Carl Jung* VIII.vii. East Sussex: Routledge. p. 3391.
†Main, R. (1997). *Jung on Synchronicity and the Paranormal*. Princeton University Press. p. 1.

21. Ibid p. 129.
22. Ibid p. 162.
23. Ibid p. 152.
24. Bell, J. (1987). *Speakable and unspeakable in quantum mechanics*. Cambridge University Press. p. 170.
25. Bohr, N. Discussion with Einstein on epistemological problems in atomic physics *Quantum Theory and Measurement* ed J. A. Wheeler, Princeton University Press (1983) p. 44.
26. Marin, J. M. (2009). *Eur. J. Phys.* 30, 807.
27. Schrödinger. (2012). p. 1.
28. Einstein A. (1981). *Albert Einstein, The Human Side*. Princeton University Press. p. 40.
29. Marin (2009).
30. Jammer, M. (1999). *Einstein and Religion: Physics and Theology*. Princeton University Press. p. 125.
31. Marin, p. 815.

Chapter 4: How Weird Is It?

1. Verification of probability inferences requires a large number of trials. To verify that the probability of getting a head in tossing a coin is 50%, you have to toss a coin many many times, or toss many many coins once. Only then will you see that the number of heads showing up is about half the total.
2. To answer the question, we need to know the ratio of black to white paint. There is half a gallon plus one drop of black and exactly half a gallon of white. There are approximately 100,000 drops in a gallon. Out of these, 50,000 are black and 50,000 are white when the paint is perfect gray. Adding one black drop changes the ratio to 50,001/50,000 = 1.00002. Going back to the coins, we conclude that, for our instrument to be able to distinguish the shade of black, the ratio of the number of black coins to white coins must be 1.00002. Since the total number of coins is one trillion, the actual number of black heads must be 500,005 million leaving 499,995 million white tails. (The ratio of these two numbers is easily seen to be very nearly 1.00002.)
3. This number is obtained using the formula given in Appendix, or more conveniently, an equivalent formula involving an exponential function which is easier to use.
4. To go from probability to "odds are one in ..." you take the reciprocal of the probability. This comes from the formula for the number of successes on page 166 in which the number of successes is one on the left-hand side.
5. It is exponentially beyond our ability to see each individual act of the salt shaker. We can detect the sequence of events only if they occur less than 10–12 times per second. All animations use this disability of our sight and show more

184 Notes

than 18 frames per second, fooling us into believing that events are occurring continuously and smoothly.
6. By the formula on page 166.
7. https://www.nobelprize.org/prizes/physics/1918/planck/lecture/.
8. Systems consisting of a few atoms can violate the second law of thermodynamics, but, ordinarily, we never deal with a few atoms.
9. Quantum dots are man-made constructs with an electron confined in a (potential) well … a quantum jar-candy system.

Chapter 5: From Duality to Mysticism

1. Photoelectric effect was known since the late nineteenth century, but no explanation existed before 1905, when Einstein proposed the particle nature of EMWs. He explained the effect by suggesting that the interaction of photons with the electrons in the metal releases the electrons, creating an electric current. Curiously, Einstein won the Nobel Prize for photoelectric effect, but not for his special and general relativity theories, both of which are far more significant and have brought Nobel Prizes to many physicists who have contributed to those fields.
2. The so-called *linear* momentum is the product of mass and velocity. *Angular momentum* applies to objects that move around a center. The product of linear momentum and the distance from the center defines the angular momentum of an object. Planets have angular momentum with the center being the sun. A bicycle wheel has angular momentum because the rim moves around the shaft (the center). The fact that planets speed up (their linear momentum increases) when they get close to the sun (their distance from the center decreases) is the result of the *conservation of angular momentum*, which states that the product of distance and linear momentum does not change during the motion of the object. The fact that your moving bicycle doesn't fall as you ride on it is also a consequence of angular momentum conservation. A force (more accurately, a torque) is needed to change the angular momentum of a moving bicycle (tip it sideways), just as a force (friction of the brakes, for example) is needed to change the linear momentum (or the speed) of the bicycle.
3. It turns out that hydrogen energies are *negative*, as are the energies of all atoms. So larger energies correspond to smaller absolute values.
4. The color of light is determined by either its frequency or its wavelength. The two are uniquely related to each other because the product of frequency and wavelength gives the unique constant value of the speed of light.
5. If you pass the light emitted by an incandescent light bulb through a prism, you'll see a continuous spread of light of various wavelengths (colors) from red to violet, the rainbow colors. If you pass the light emitted from heated elements like hydrogen, helium, or neon, you'll see only discrete bands of

colors, the *spectral lines* of that element. Each element has its own unique spectral lines acting as its fingerprint. Astronomers have been able to identify the constituents of stars and galaxies by looking at these lines.
6. There were other serious problems with the model, not the least of which was lack of explanation of what keeps the electron from collapsing into the nucleus when it is on the Bohr orbit.
7. This relation was already known from the special theory of relativity.
8. To elucidate this limitation, suppose you are driving on a highway with your cruise control on, and your friend traveling with you asks, "What is the speed of the car?" You look at the speedometer and respond, "65 mph." Your friend says, "Is it exactly 65?" You look at the speedometer again and say, "It is actually 66." Your friend says, "Are you sure? Couldn't it be 65.5 mph?" Even if you affirm your friend's suggestion, he may demand more accuracy: 66.45 or 66.55 or …. You can never tell your friend the *exact* reading of the speedometer. In this case the spread, or uncertainty, may be 0.1 mph—from 66.45 to 66.55. This information is usually written as "The speed of the car is 66.5 ± 0.05 mph."
9. In the most accurate measurements, a typical product of uncertainties in momentum and position for a bullet is about a millionth in scientific units. That is 1,000,…,000 (replace the dots with 22 zeros) times larger than the Planck constant! It is safe to say that ordinary objects *never* violate the uncertainty principle.
10. This doesn't mean that quantum physics is irrelevant to the *atoms and molecules* composing everyday objects. The entire field of condensed matter physics applies quantum physics to the bulk matter to explain such properties as electric and heat conductivity (see page 128). The irrelevance of quantum physics concerns the object as a whole.
11. https://www.bbc.com/news/world-asia-india-46778879.
12. For the purposes of this experiment, ignore gravity and assume that the trajectory of bullets are perfectly straight lines.
13. Actually, the blob of photons is slightly different from that of bullets, but we need not dwell on this difference.
14. See page 84.
15. Photons do not obey the Schrödinger equation, which is suitable only for non-relativistic particles. Photon is the epitome of a relativistic particle. For the case of a double-slit experiment, however, the wave function of a photon turns out to be the same as a non-relativistic particle.
16. If you are familiar with optics, you may note that this statement is incomplete. The shape of the blob is actually that of *diffraction*: there is a big blob at the center, but there are two "secondary" weak blobs on either side of the big one

that are separated from the central blob by dark regions. There are also tertiary and higher order blobs, but they are so weak that they can hardly be seen.

17. For the interested reader, the Appendix contains a slightly more technical (and more convincing) discussion.
18. By the word "understand" Feynman is expressing the urge to comprehend quantum mechanics in terms of concepts expressed in ordinary human languages; an urge that has to be avoided but, unfortunately, has been embraced by some quantum physicists, including the founders of quantum physics.
19. The beam of unpolarized light can have polarization in any random direction, not just horizontal and vertical. However, a general polarization can be thought of as a combination of horizontal and vertical polarizations, with a certain fraction being one or the other. We can, for instance, pass the beam through a polarizer making a 45° angle with the vertical *before* the beam reaches the slits. This way, we know that half of the photons are vertically polarized and the other half horizontally. I'm ignoring this technical detail here.
20. The word "product" used here is not the same as what you get when you multiply two numbers. The word refers to a generalized notion of product used in higher mathematics. To see that the notion is not the same as that used in ordinary multiplication, suffice it to say that it is possible for the "product" to be zero with neither of the factors being zero! If you are familiar with "dot product," you know that the dot product of two perpendicular *nonzero* vectors is zero.
21. Marin, J. M. (2009). *Eur. J. Phys.* 30, 807.
22. Here I have to distinguish between classical and quantum randomness. Temperature is an example of classical randomness. It is related to the average kinetic energy of the molecules in the boiling water or steam. As long as the sample contains billions and billions of molecules, the value of the temperature is exact. This is due to the *law of large numbers* in probability theory (an example of which was the salt shaker full of microscopic painted coins discussed on page 39), which states that when the number of objects in a random process tends to infinity, the fluctuation (or uncertainty) in the average tends to zero. For a sample consisting of a small number of molecules, however, temperature exhibits large fluctuations. In particular, the notion of temperature does not even exist for a single molecule. By contrast, the notion of position—or momentum, or energy, or any other particle property—exists for a single quantum particle like electron, but it obeys probabilistic laws.
23. The total number of possibilities is 16. To verify this, align the coins horizontally with the first coin on the left and the fourth coin on the right and the second and third coins between the two. Then you have these sixteen arrangements: HHHH, THHH, HTHH, HHTH, HHHT, TTHH, THTH, THHT, HTTH, HTHT, HHTT, TTTH, TTHT, THTT, HTTT, TTTT. As you can see, only six of these have two heads. So, the probability

of getting two heads when tossing four coins—or the fraction of times two heads show up—is 6/16 = 0.375.
24. Recall the trillion microscopic coins and the salt shaker experiment on page 39. We had to wait more than the age of the universe to see anything but perfect gray.

Chapter 6: Quantum Consciousness Crosses the Atlantic

1. Critchfield, C. (1996). "The Oppenheimer I Knew," in *Behind Tall Fences: Stories and Experiences about Los Alamos at Its Beginning*, 169–77 (Los Alamos, NM: Los Alamos Historical Society).
2. Johnson, G. (1999). *Strange Beauty: Murray Gell-Mann and the Revolution in Twentieth-Century Physics*. Alfred A. Knopf. p. 80.
3. The New York Review of Books, July 19, 2007; https://bit.ly/3SreAzj; https://bit.ly/3DGEdb6; https://bit.ly/3e3TsQY.
4. The "design argument" is a reference to Intelligent Design, a religious alternative to the scientific theory of evolution. When the term "scientific creationism" was unequivocally exposed as an effort to repudiate evolution in favor of the Biblical story of creation, its adherents found "Intelligent Design" a more slyly appealing sales pitch. The design argument actually goes back to the eighteenth-century theologian William Paley, who began the first chapter of his famous treatise, *Natural Theology – or Evidence of the Existence and Attributes of the Deity*, with "In crossing a heath, suppose I pitched my foot against a *stone*, and were asked how the stone came to be there, I might possibly answer, that, for any thing I knew to the contrary, it had lain there forever: nor would it perhaps be very easy to shew the absurdity of this answer. But suppose I had found a *watch* upon the ground, and it should be enquired how the watch happened to be in that place, I should hardly think of the answer which I had before given, that, for any thing I knew, the watch might have always been there." By contrasting a naturally occurring stone with a man-made watch, Paley is hinting at a *creator*, as he concludes "that the watch must have had a maker: that there must have existed at some time, and at some place or other, an artificer or artificers, who formed it for the purpose which we find it actually to answer; who comprehended its construction, and designed its use."
5. Dyson, F. J. (1988). *Infinite in All Directions*. Harper and Row. p. 297.
6. Ehrenreich, B. (2010). *Bright-Sided: How Positive Thinking is Undermining America*. Picador. p. 166.
7. Wigner, E. P. (1967). *Symmetries and Reflections*, Indiana University Press. p. 172.

8. See Chap. 7.
9. Weyl, H. (1952). *Space, Time, and Matter*. Dover. p. 4.
10. Marin, J. M. (2009). *Eur. J. Phys.* 30, 807.
11. As a first-year graduate student of the Physics Department in 1975, I had the privilege of receiving Wheeler's guidance and encouragement at the start of my professional career.
12. Barrow, J.D, Davis, P. C. W. and Harper, Jr. C. L. (eds.) (2004). *Science and Ultimate Reality: Quantum Theory, Cosmology, and Complexity*, Cambridge University Press. p. xviii.
13. Wheeler, J. A. (1989) in Proc. 3rd Int. Symp. *Foundations of Quantum Mechanics*. Tokyo. pp. 354–368.
14. Ibid.
15. See page 68.
16. Barrow, J. and Frank J. Tipler, F. (1988). *The Anthropic Cosmological Principle*. Oxford University Press. p. vii.
17. Danesh, H. B. (2000). *The Psychology of Spirituality*. Sterling Press. p. 36.
18. Dossey, L. (2008). *Explore* **4**(6), 341.
19. Beauregard, M., Schwartz, G. E., Miller, L., Dossey, L., Moreira-Almeida, A., (2014). *Explore* **10**(5), 272.
20. Dyson, F. *Infinite in All Directions* Harper and Row, New York (1988) p. 298.
21. Horgan, J. *The End of Science* (1997), Broadway Books, New York, p. 255.
22. Esfeld, M. (1999). Essay Review: Wigner's View of Physical Reality, published in *Studies in History and Philosophy of Modern Physics*, 30B, pp. 145–154, Elsevier Science Ltd.
23. Marin, p. 813.

Chapter 7: Eastern Plague of the Sixties

1. Despite their heavenly appearance, gurus were notoriously worldly among their inner circles. Sex, violence, and abuse were common themes in many, if not all, ashrams founded by the immigrant gurus. An encyclopedic and documented volume narrating the true nature of the gurus and swamis is Falk, G. (2009). *Stripping the Gurus: Sex, Violence, Abuse and Enlightenment*. Million Monkeys Press.
2. Nuclear *physics* is the neutral science behind the *beneficial* technology that goes into the construction of an MRI machine and the *destructive* technology that goes into the manufacture of a nuclear bomb.
3. See http://bit.ly/2uq9eey and http://bit.ly/2JHPL3w.
4. http://bit.ly/2YDfhui.
5. https://en.wikipedia.org/wiki/Fundamental_Fysiks_Group.
6. http://bit.ly/2JGMHEM.

7. http://stardrive.org/index.php.
8. http://bit.ly/2U1apAr.
9. http://bit.ly/2TCEQYK.
10. Capra, F. (2010). *The Tao of Physics: An Exploration of the Parallels between Modern Physics and Eastern Mysticism*. Shambala Publications Inc. p. 11.
11. Ibid. pp. 26–29.
12. Zukav, G. (1980). *The Dancing Wu Li Masters*. Bantam Books. p. xxvii.
13. See page 26 for a connection between Pauli's mysticism and Jung's.
14. Ibid. p. 29. Recall that "consciousness" is a spirit, force, presence, or psyche that blankets the entire universe, not the trait of a conscious person.
15. Ibid. pp. 62–63.
16. Ibid p. 5.
17. When a person on a train that moves at 65 mph throws a tennis ball forward at 25 mph, a person on the ground measures the speed of the tennis ball to be $65 + 25 = 90$ mph. This is called the *Law of Addition of Velocities*. When a person on a spaceship that moves at 299,000 km/s throws a beam of light forward with a laser gun at the speed of 300,000 km/s (the universal speed predicted by Maxwell), a person on the ground measures the speed of the laser beam to be—not $300,000 + 299,000 = 599,000$ km/s, but—300,000 km/s. The motion of the spaceship does not affect the speed of light as the motion of the train affects the speed of a tennis ball. Because c is distance divided by time, if c doesn't change with motion, then both distance and time must change in such a way that their ratio remains constant. A quantitative analysis of this last statement leads to the relativistic notions of *length contraction* and *time dilation*.
18. Zukav, pp. 154–155.
19. Zajonc, A. (2004). *The New Physics and Cosmology: Dialogues with the Dalai Lama*. Oxford University Press. p. 205.
20. To see this, consider an observer for whom the electron and positron, once produced, move with the same momentum in opposite directions (such an observer always exists). For this observer, the total momentum of the pair is zero. But the momentum of the initial photon can never be zero because it always moves (at the speed of light, by definition). If one observer cannot see the creation of an electron-positron pair out of a single photon, no observer can. On the other hand, *two* photons moving in opposite directions and having sufficient momenta can collide and produce an electron-positron pair. In fact, this process occurred frequently during the first three minutes after the big bang.
21. Zukav, p. 205.
22. See page 172.
23. Hadrons and quarks are further discussed in Chaps. 9 and 10.

24. Capra, F. (1974). *Am. J. Phys.* **42**, 15.
25. Capra, *The Tao of Physics*, p. 290.
26. Zukav, p. 314.
27. https://www.nobelprize.org/prizes/physics/2004/gross/facts/.
28. Ibid p. 36.

Chapter 8: The "Quantum" Healer

1. Falk, G. D. (2009). *Stripping The Gurus: Sex, Violence, Abuse and Enlightenment*. Million Monkeys Press. pp. 58–68.
2. I haven't been able to get a hold of the 15th printing. So, I don't know which "edition" it belongs to.
3. https://bit.ly/3Tu5X9n.
4. For the benefit of other medical doctors who want to follow Chopra's treatment, shouldn't he talk about how to measure consciousness? After all, don't you need to measure consciousness to find a "jump" in it? But he doesn't. He can't. Because consciousness, as conceived by the Eastern theosophy, is not measurable, detectable, touchable, or observable.
5. Chopra, D. (1989). *Quantum Healing*. Bantam Books. pp. 15–16.
6. See page 51.
7. Ibid, p. 96.
8. Chopra, D. and Kafatos, M. (2017). *You are the Universe: Discovering Your Cosmic Self and Why It Matters*. Harmony.
9. Ibid. p. 45.
10. See page 43.
11. See pages 68 and 72 for a critical analysis of the anthropic principle.
12. Ibid. p. 61.
13. Ibid. pp. 92–95.
14. See the end of Chap. 7 and the beginning of this chapter.
15. Ibid. p. 104.
16. Baryons are discussed in Chap. 9. The 4% mentioned here are almost exclusively hydrogen and helium, whose masses—thus, energies—are almost entirely concentrated in their nuclei, which are composed of protons and neutrons. And protons and neutrons turn out to be baryons.
17. https://bit.ly/3AyGEqS.
18. Ibid. pp. 111–112.
19. Professors at prestigious institutions of higher learning can have opinions and philosophies as ludicrous as the most notorious quacks. Even the Nobel Prize does not solidify the rationality of scientists on matters of philosophy and religion. Recall that even the founders of quantum physics were not immune from quackery.
20. Ibid. pp. 155–160.

21. Ibid. pp 241–245.
22. For more on the notion of "mainstream" and why revolutionaries like Einstein and Planck are also mainstreamers, see Hassani, S. (2015). "Postmaterialist Science? A Smokescreen for Woo", *Skeptical Inquirer* **39**, No. 5, 38–41. (Online at http://bit.ly/2mksm8Q).
23. Chopra and Kafatos, p. 214.
24. Ibid. pp. 237–239.
25. https://www.youtube.com/watch?v=o-ijyqWzDrY minute marks 0:01 to 0:15.
26. http://bit.ly/2HmGSey.
27. Because energy is a *property* of material objects, and properties themselves are non-material, New Agers take $E = mc^2$ to mean the equivalence of soul or consciousness (the left side of the equation) with matter, the m on the right side. The fallacy of this reasoning becomes evident when we assert that the redness of an apple or blackness of coffee are non-material. There is no redness or blackness without a material object that carries the color. Likewise, since E is a property, the left-side of $E = mc^2$ is *always* the energy carried by some material objects. See also page 89.
28. The emphasis on "terrestrial" is meant to remind the reader of the importance of gravity, which by 1915 crystalized in Einstein's general relativity, one of whose predictions was the Big Bang as the start of the universe itself. Gravity, however, had insignificant role in the study of terrestrial matter.
29. Mayr, E. (1982). *The Growth of Biological Thought*, Harvard University Press. pp. 55–56.
30. To appreciate the difficulty of the task confronting molecular biologists, consider a water molecule which contains only three atoms, two hydrogen and one oxygen. We have known water ever since our existence as a curious species. And even though chemists have studied water molecules for centuries and know a tremendous amount of its physical and chemical properties, they don't understand it fully. DNA was discovered less than seventy years ago and is known to have over three hundred billion atoms! To understand how it works—including its duplication capability and its "genetic program"—may require decades, if not centuries. However instead of waiting and giving science a chance, new vitalists propose a qualitative difference between biology and the physical sciences and proclaim the futility of the goal of understanding DNA in terms of physical laws in the future.
31. My emphasis on elemental parts does not invalidate fundamental concepts such as entropy, which, by its very nature requires composite systems. Nevertheless, one cannot ignore the importance of "elemental parts" in the unlocking of the puzzling behavior of entropy. (See page 43.)
32. The number of neurons by itself is not a good indicator of intelligence. Our brain holds almost 90 billion neurons, while an African elephant has almost three times that many neurons in its brain. The additional marker of

intelligence is the intricacy of the communication between neurons, i.e., the number of neurons with which a single neuron communicates.

33. Sagan, C. (1977). *Dragons of Eden*. Random House. pp. 246–247.
34. https://www.youtube.com/watch?v=o-ijyqWzDrY minute marks 1:17:12 to 1:17:28.

Chapter 9: Basic Building Blocks

1. By the beginning of the twentieth century, three "rays" were identified as the by-products of nuclear radioactivity: alpha rays, beta rays, and gamma rays. It was discovered later that these rays consisted of particles. The nucleus of helium composed alpha rays, electrons and neutrinos composed beta rays, and gamma rays were identified as highly energetic photons.
2. These are: PE_{n1}, between nucleus and first electron; PE_{n2}, between nucleus and second electron; and PE_{12}, between the two electrons.
3. A deuteron is the nucleus of deuterium—or heavy hydrogen. A *heavy water* molecule consists of two deuterium atoms and one oxygen atom.
4. See page 44 for a discussion of tunneling. In the example of a jar-candy system, the wall of the jar is a potential barrier.
5. Maxwell predicted the speed of the EMWs, including light, to be 300,000 km/s *in vacuum* (see page 189). In matter, light slows down. For example, in water, it moves at about 255,000 km/s, and in glass at 200,000 km/s. The interior of a star is so dense that, even though a typical star like the Sun has a radius of about 700,000 km—and were it vacuum, light would cover it in a little over two seconds—it takes light millions of years to cover the radius.
6. For example, the orderly packed atoms in a crystal consisting of an almost infinite identical unit cells, essentially reduces the study of the entire crystal to that of a single unit cell consisting of only a few atoms.
7. Reductionism holds even at the level of classical equations of motion. There is no such thing as the temperature or entropy of a single particle. However, if you put many such particles together to form a gas and apply the classical equation of motion, which is the sum of its parts,* the notions of temperature and entropy transpire.
8. The two functions turned out to be associated with the *spin* of the electron, which could have two states: *up* or *down*. The Dirac equation incorporated the spin of the electron automatically. This in itself was a monumental

---

*The left side of the equation of motion is the sum of the products of masses and accelerations of the particles; the right side is the sum of the forces experienced by each pair of particles in the system.

accomplishment because, up to that point, spin was an unsolved quantum mystery. Further discussion of spin can be found in Appendix.

9. Pure energy is a misleading misnomer often exploited by mystics to claim that matter turns into pure energy—of the sort, of which soul and consciousness are made up. What really happens is that the electron and positron turn into two photons carrying the energy.
10. Capra, pp. 76–81.
11. https://bit.ly/3XjK3bX.
12. See page 171 for a more elaborate discussion of spin.
13. See page 171.
14. Ne'Eman, Y. and Kirsh, Y. (1989). *The Particle Hunters*, Cambridge University Press. pp. 200–201.
15. A phenomenological theory—as opposed to a fundamental theory—is a theory that is most directly connected to observation. It is designed to simply explain (or summarize, usually in mathematical formulas) what experiments and observations reveal.
16. Regrettably, some critics of science with holistic views—who, oddly, are populated in the community of philosophers and sociologists of science—purposefully use the word "fundamentalism" as a synonym for reductionism to impart to it the negative political and religious connotation that is carried by the word.
17. Capra. p. 30.
18. Neutrinos are so elusive that to stop half of the ones produced in a typical decay, one needs a slab of lead that is, … not a few centimeters, or a few meters, or a few kilometers, but a few *light years* thick!

Chapter 10: The Standard Model

1. Massless particles *must* move at the speed of light. This can be seen from the relativistic equation of the energy of a particle of mass m moving with velocity v: $E = \frac{mc^2}{\sqrt{1-(v/c)^2}}$, where c is the speed of light. If an object moves with the speed of light, i.e., if $v = c$, then the denominator becomes zero, and, if m is not zero, E becomes infinite. So, the only way that E does not go to infinity is for m to be zero. Now we can see why massless implies long range forces: since the particle moves at light speed, it can go a long distance in a very short time.
2. If you place a ball at the side inside a bowl, it will slide down and oscillate for a while, but eventually comes to rest at the bottom of the bowl. The points on the side have a higher (potential) energy; the bottom of the bowl has the lowest energy. Therefore, it is the most stable of all points: while other points

cannot hold the ball, i.e., the ball is unstable at those points, the bottom holds the ball forever. Think of the vacuum as the bottom of a bowl.
3. https://pdg.lbl.gov/2023/download/db2022.pdf.
4. Naked, because the "clothed" charmonium has an anti-charm quark which cancels the "charmness" property.
5. They chose τ from *tritos*, Greek for "third," the third charged lepton.
6. Bona fide is the word. In 1989 a couple of scientists hastily announced their experimental result of "cold fusion," not to the physics community which could evaluate the validity of their claim, but to the news media. After more than a third of a century, the cold fusion result has not been independently verified, and aside from a cult of believers, no physicist takes cold fusion seriously.
7. The name "color" has no relation to the kind of stimulus to which our eyes are sensitive. In fact, although the three colors associated with quarks are usually called red, green, and blue, when describing the colors mathematically, one simply uses 1, 2, and 3.
8. Gravity is not included in these interactions. It plays no role in the energies and distances available in accelerators. However, it becomes important in energies that were once available moments after the big bang.
9. Capra, pp. 316–317.
10. Ibid, p. 317.

Chapter 11: Epilogue

1. Merton, R. K. (1973). *The Sociology of Science: Theoretical and Empirical Investigations*. University of Chicago Press. pp. 371–383.
2. "Peer review," a phrase that tacitly describes articles published in real scientific journals, has recently been hijacked by pseudoscientists, who repeatedly mention it like a mantra to validate their publications in journals edited, associate edited, and "peer reviewed" by pseudoscientists. (See page 74.) A real prestigious scientific journal does not have to advertise its peer review process to gain credibility. The process is given, and the journal is credible because of the rigor of the process.
3. See page 70.
4. The very publication of the periodical, *Explore*, by the reputable publisher Elsevier points to an infiltration of pseudoscience into academia; and the immediate rise of books like *You are the Universe* to bestselling status signals the dangerous level to which pseudoscience has become acceptable in our society.
5. See page 26.
6. As is sometimes the case, physicists stumble upon some mathematical ideas that had existed, but gone unnoticed. Spin geometry is related to *Clifford algebra*, a rudimentary form of which was the brainchild of William Clifford, a

nineteenth-century English mathematician. Spin geometry is one of the latest examples of the kinship between mathematics and physics discussed earlier (see page 131).
7. The final product of matter-antimatter annihilation is not pure energy. It is a pair of photons (light particles) which *carry* the energy associated with the initial mass—via $E = mc^2$—of the amount of matter and antimatter present. See page 88 for more details.
8. Heisenberg, *Physics and Beyond*, p. 85.
9. See pages 43 and 56.
10. Bell, J. (1987). *Speakable and unspeakable in quantum mechanics*. Cambridge University Press. p. 170.
11. As later articulated by Zukav, p. xxvii.

Appendix

1. https://en.wikipedia.org/wiki/Homeopathy.
2. A full circle subtends 2π = 6.283185307 rad, i.e., 360° equals 6.283185307 rad, or each radian is 57.296°.
3. For example, $3x - 4$ is a polynomial of degree one and $2x^2 - 8x + 6$ is a polynomial of degree two. The first-degree equation $3x - 4 = 0$ has a single solution, $x = \frac{4}{3}$, and the second-degree equation $2x^2 - 8x + 6 = 0$ has two solutions, $x_1 = \frac{8+\sqrt{8^2-4\cdot 2\cdot 6}}{4} = \frac{8+\sqrt{16}}{4} = 3$ and $x_2 = \frac{8-\sqrt{8^2-4\cdot 2\cdot 6}}{4} = \frac{8-\sqrt{16}}{4} = 1$.
4. Contrary to, for example, Pythagoras' theorem which evolved out of Egyptians trying to create a right angle when parceling lands.
5. Think of spin as a top whose axis of rotation need not be vertical. The direction of the spin of an ordinary top can smoothly vary between vertical and some maximum value. The direction of a quantum spin is quantized.
6. There is a fundamental difference between Mendeleev periodic table of elements and the eightfold way: The former is purely empirical, the latter purely mathematical.
7. Of course, the electron can remain an electron and emit a gauge boson as the first diagram on page 175 demonstrated. Similarly with muon.

Bibliography

Armstrong K (1993) A history of god. Alfred A. Knopf
Bell J (1987) Speakable and unspeakable in quantum mechanics. Cambridge University Press
Bohr N (2010) Atomic physics and human knowledge. Dover Publications
Byrne R (2006) The secret. Atria Books
Capra F (2010) The Tao of physics: an exploration of the parallels between modern physics and eastern mysticism. Shambala Publications
Chopra D (1989) Quantum healing. Bantam Books
Chopra D, Kafatos M (2017) You are the universe: discovering your cosmic self and why it matters. Harmony
Clark E (1953) Indian legends of the Pacific northwest. University of California Press
Dyson FJ (1988) Infinite in all directions. Harper and Row
Eddington A (1920) Space time and gravitation. Cambridge University Press
Ehrenreich B (2009) Bright-sided: how positive thinking is undermining America. Picador
Einstein A (1981) Albert Einstein, the human side. Princeton University Press
Enz CP (2002) No time to be brief: A scientific biography of Wolfgang Pauli. Oxford University Press
Falk G (2009) Stripping the Gurus: sex, violence, abuse and enlightenment. Million Monkeys Press
Heisenberg W (1958) Physics and philosophy: the revolution in modern science. Harper and Row Publishers
Heisenberg W (1972) Physics and beyond: encounters and conversations. Harper and Row Publishers
Hofstadter R (1970) Anti-intellectualism in American life. Alfred A. Knopf
Horgan J (1997) The end of science. Broadway Books

Jammer M (1999) Einstein and religion: physics and theology. Princeton University Press
Johnson G (1999) Strange beauty: Murray Gell-Mann and the revolution in twentieth-century physics. Alfred A. Knopf
Lachapelle S (2011) Investigating the supernatural: from spiritism and occultism to psychical research and metapsychics in France, 1853–1931. Johns Hopkins University Press
Mayr E (1982) The growth of biological thought. Harvard University Press
Merton RK (1973) The sociology of science: theoretical and empirical investigations. University of Chicago Press
Müller M (1884) The Upanishads, Part 2. Oxford at the Clarendon Press
Ne'Eman Y, Kirsh Y (1989) The particle hunters. Cambridge University Press
Pauli W (1994) Writings on physics and philosophy. Springer
Pellegrino C (1994) Return to Sodom and Gomorrah. Random House, New York
Plato (1987) The Republic. Penguin Books
Quinn S (1995) Marie Curie: a life. Simon and Schuster
Sagan C (1977) Dragons of Eden. Random House
Schopenhauer A (2007) The world as will and representation. Routledge
Schrödinger E (1992, 2013) What is Life? with mind and matter & autobiographical sketches. Cambridge University Press
Weyl H (1952) Space, time, and matter. Dover
Wigner EP (1967) Symmetries and reflections. Indiana University Press
Zukav G (1980) The dancing Wu Li masters. Bantam Books

Index

A
Above World, 10
accelerator, 129, 130, 133
acupuncture, 3, 181
　at Harvard, Yale, Mayo Clinic, 6
　NCCIH, 6
adrenaline, 101
Akhenaten, 15
Alexandria
　Museum, 13
al-Khwarizmi, Muhammad ibn Musa
　and algebra, 170
　and algorithm, 170
alpha particle, 50, 127, 129
alternative fact, 3
alternative medicine, 3, 5, 6, 97, 99, 116
　at Harvard, Yale, Mayo Clinic, 6
　probability, 41
alt-right, 4
American Cancer Society, 41
Anaxagoras, 14
Anderson, Carl
　discovers positron, 130
angular momentum, 50
Annalen der Physik, 113
Anquetil du Perron, Abraham Hyacinthe, 17

Anthropic Principle, 105
　and Wheeler, John, 72
anti-charm, 146
anti-electron
　same as positron, 130
anti-matter, 130
anti-proton, 130
anti-vaccination, 3, 6
anti-vaxxer, xii, xiii
Anu, 15
Archimedes, 126
Aristarchus of Samos, 13
　reductionism
　　heliocentrism, 126
Aristotle, 14, 23, 119, 120
　Lyceum, 13
　Physics, 132
　　what is motion?, 119
Aten, 15
Athena, 12
Atman, 19, 47
atom
　movie made of, 45
　moving it, 45
　plum-pudding model, 50
　seeing it, 45

Index

atomic nucleus, 50, 122, 127–129, 148
  and electric force, 149
  outcome of reductionism, 127
  protons and neutrons, 149
  radioactive, 177
atomism, 180
axon, 102
Ayurveda, xiii, 2, 3, 99–101, 116
  harm of, 4
  at Harvard, Yale, Mayo Clinic, 6
  NCCIH, 6

B

Babylonian temples
  centers for astronomy, 11
background radiation, 106
Bannon, Steve, 3
Barrow, John
  Anthropic Principle
    Templeton Prize, 68
baryon octet, 173
baryonic charge
  its conservation, 134
baryons, 133
basic building blocks
  Nobel Prizes given for, 152–154
Beatles, 2
Beethoven, 155
Bell's inequality, 161
  no EPR paradox, 46
  and non-locality, 46
Bell, John, 19, 46, 154, 161, 183, 195
  on Anthropic Principle, 160
  on conscious observer, 28
  on Eastern theosophy, 160
  and EPR paradox, 46
  on information, 160
  on observer-created reality, 160
  opposes theosophy, 31
  and Nobel Prize, 161
  non-locality, 180
Below World, 10

Bevatron, 130
  and anti-proton, 130
Bhagavad Gita
  and Oppenheimer, 66
Bible, 116
big bang
  creates time, 105
  and standard model, 150
biology
  and reductionism, 120, 127
Bishop Berkeley
  and Schrödinger, 28
black body radiation curve, 87, 113
  Planck formula
    a phenomenological theory, 136
black hole, 71
Bohr model
  of H-atom, 51
Bohr, Niels, 7, 24, 30, 31, 50–53, 61,
    63, 65, 69, 71, 108, 115, 158,
    159, 181, 183
  admission of guilt, 29
  complementarity
    consciousness, 33
  Eastern mysticism, 24
  Einstein on, 32
  H-atom
    spectral lines, 51
  and John Wheeler, 71
  and objectivity, 61
  Post-Materialist Science
    manifesto, 75
Bohr orbit, 51
Bolshevik revolution, 23
Boltzmann, Ludwig
  atoms
    probability, 42
  makes sense of entropy, 43
  praised by Planck, 42
  statistical mechanics, 43
  suicide 1906, 42
bootstrap hypothesis, 137
  a defunct hypothesis, 92–94

and Eastern theosophy, 93
Eastern thought, 151
and heaven of Indra, 93
holistic theory, 150
Born, Max, 157
and Oppenheimer, 65
probability amplitude, 44, 47
boson
    gauge, 141
    Goldstone, 144
    Higgs, 147–149
bottom quark, 147
Brahmajala Sutta, 16
Brahman, 15, 19, 47
Brahmin, 16
brain
    reticular theory, 122
Brookhaven National Laboratory, 135, 146
Brownian motion, 42
Buddha, 17, 19, 24
    against astronomy, 16
Buddhism, 2, 19, 24, 25, 70, 86, 94
Byrne, Rhonda, 1

C

Cajal, Santiago Ramon y, 122
calculus
    Leibniz, 156
    Newton, 156
calorie, 92
Caltech, 136
Canaan, 11, 15
canine mysticism, 94–95
capitalism, 23
Capra, Fritjof, 2, 93–95, 112, 137, 151, 152, 190
    denies QP-relativity unification, 131
    denies success of GTR, 151
    ignoring QCD's success, 150
    and Lord of Dancers, 81
    quark
        spiritual?, 151
        "third stage", 139
Carnot, Sadi, 17
CERN, see European Center for Particle Physics, 145
chakra, 3
    at Harvard, Yale, Mayo Clinic, 6
charm quark, 146
    discovered, 146
    naked, 146
charmonium, 146
chemistry
    and reductionism, 127
Chew, Geoffrey
    and bootstrap, 92
    and mysticism, 93
Chicken Soup for the Soul
    and quantum physics, 1
Chinese Whisper, 180
    and myths, 10
Chopra, Deepak, xii, xiii, 2, 27, 80, 97, 98, 106, 107, 109, 110, 112, 114, 116, 117, 190
    admiring Maharishi, 97
    and Schrödinger, 27
    attacking physics, 116–118
    modern St. Augustine, 123–124
    and professional honesty, 99
    prophet of alternative medicine, 116
    QP-mysticism connection, 123
    and Schrödinger, 27
    steals neuropeptides, 101
    vision of Maharishi, 98
    You are the Universe, 103
Chopra's quantum mechanics
    how sneeze causes earthquake, 102
Civil Rights Movement, 79
classical physics
    as mechanistic, 86
    begets modern physics, 87–88
Clifford, William, 194
climate change, xi, 3, 67
    denial, 6

cold fusion, 193
color charge, 149
combustion, 91
communicating coins?, 40
complementarity principle, 24, 26, 53–54, 59
   Buddhism and Hinduism, 25
   and consciousness, 30, 61
   and double-slit experiment, 58
   and duality, 58
   and observer-created reality, 54
   from uncertainty principle, 52
   no such thing, 60
condensed matter physics, 128
Confessions
   manifesto of Dark Ages, 123
confinement, 143, 149
conscious coin?, 39–40, 86
   absurdity of, 47
conscious universe?, 108
consciousness, 2, 15, 16, 18, 21, 22, 27, 28, 30, 32, 33, 40, 47, 56, 58, 61, 62, 68–70, 72–77, 80, 84, 85, 100, 101, 108–112, 114–116, 118, 121, 123, 159–161, 189–191, 193
   and Ayurveda, 99
   as creator of reality, 19
   creates everything, 109
   creates life, 109
   creates opposites, 110
   creates time, 109
   disproved by probability, 39
   God in disguise, 110–111
   jumping, 99–100
   and non-locality, 18
   predates big bang, 109
   property of brain, 123
   qualia, 109
      composition of universe, 109
      quantum domain, 109
   in quantum physics, 8
   what is it?, 121–123
   where it comes from, 119–123
conservation
   baryonic charge, 134
   electric charge, 134
conspiracy, 3, 4, 6, 13, 108
conspirituality, xiii, 4
correspondence principle, 88
Coulomb
   Charles-Augustin de, 174
Coulomb potential barrier, 128
COVID, xi, xii, 6
critical thinking, ix, xii, 12, 13
Curie, Marie, 21
Curie, Pierre, 21
cyclotron, 129

D

Dancing Shiva, 81–84
Dancing Wu Li Masters, 2
Dark Ages, xiii, 6, 106, 108, 124
dark energy, 106
dark labyrinth
   Aristotle's physics, 131
dark matter, 106
Darwin, Charles, 156
daughter of the sky, 9
de Broglie, Louis, 51, 52, 54
Democritus, 14
determinism, 24, 43, 44, 47, 62, 181
   and objectivity, 62
deuterium, 192
deuteron, 128
diffraction, 185
Dirac equation, 130
Dirac, Paul
   QP-relativity unification, 158
   on religion, 159
DNA, 7, 120, 121, 191
   in anthropology, 121
   in archeology, 121
   double-helix, 121

in forensics, 121
in genetic engineering, 121
outcome of reductionism, 127
double-slit, 52
double-slit experiment, 55–58
and conscious photon, 84
and determinism, 58
explanation, 56–57
with polarizers, 59
doublet
quark, 146
down quark, 136
Doyle, Arthur Conan
a spiritualist, 20
dreams
probability, 40
duality, 58
and consciousness, 69
and objectivity, 61
no such thing, 60
source of mysticism, 49, 52
Dyson, Freeman, 65, 66
admission of guilt, 76
and consciousness, 66–68
and intelligent design, 67
speculates on climate change, 67
and Templeton Prize, 67

E

$E = mc^2$
and dancing energy, 88
how to abuse it, 88–91
Ea, 11
Eastern mysticism, 6, 24, 31, 62, 81, 89, 90, 138, 139, 160, 189
and classical physics, 86
and modern physics, 86
and quantum physics, 6
versus Western religion, 84
Eastern theosophy, ix, xiii, 3, 6, 7, 18–20, 22, 24–26, 31, 70, 181, 190
abhors reductionism, 93
and mathematics, 137
mixed with physics, 79
and Schrödinger, 27
Eastern thought, xii, xiii, 2, 18, 29, 82, 91, 94
unspeakable, 82
Eddington, Arthur
a mystic, 21, 22, 32, 45, 161
Einstein on, 32
Egyptian temples
centers for astronomy, 11
eightfold way, 134
a phenomenological theory, 136
prediction of eta, 173
prediction of Ω^-, 135
Einstein, ix, 7, 22–24, 26, 28–32, 42, 44, 46, 51, 69, 77, 88, 103, 105, 108, 110–112, 118, 133, 153, 155, 157–159, 161, 182–184, 191
accusing Bohr, 29
and atomic bomb, 7
on Heisenberg-Bohr religion, 32
and IAS, 66
a classical physicist, 88
explains photoelectric effect, 50
fierce anti-mysticism, 45
fighting mysticism, 7
language of Nature, 133, 135
Michele Besso, 98
and Narendra Modi, 54
opposing duality, 32
electromagnetic waves
and hot objects, 49
electron, 107
first lepton, 140
neutrino of, 140
electroweak force, 145, 177
Lie group
doublet, 149
Ellil, 15
EMW, *see* electromagnetic wave

energy
    as property of matter, 89
    what is it?, 119–120
entanglement, 18, 23
entropy, 43
Enuma Elish, 15
Epicurus, 14
    Garden, 13
EPR paradox, 46
Eratosthenes, 13
Eroica
    only one, 156
Esalen Institute, 2, 80, 81
    physics and theosophy, 81
    Zukav learns physics at, 84
ESP, *see* extrasensory perception
Euclid, 13
    Elements, 13
Euphrates, 11
European Center for Particle Physics, 148
Eusapia Palladino
    a medium, 20
evolution
    Darwin, 156
    Wallace, 156
experimentation
    is reductionist, 126
experimenter effect, 74, 108
Explore
    publishes mystical papers, 74
extrasensory perception
    and non-locality, 45

F

faith and physics, 148
faith healer
    harm of, 4
    probability, 41
Far East, 3–5, 15–17, 25, 27, 125, 181

Fasori Evangélikus Gimnázium
    Dennis Gabor, 68
    Edward Teller, 68
    Eugene Wigner, 68
    George de Hevesy, 68
    John von Neumann, 68
    Leó Szilárd, 68
    Paul Erdös, 68
    Theodore von Kármán, 68
Fauci, Dr. Anthony, 3
Fermilab
    bottom quark, 147
    top quark, 147
    Υ particle, 147
Feynman, Richard, 160
    and John Wheeler, 71
    opposes philosophy, 31
    on understanding QP, 58
field
    mathematical description of a particle, 141
    what it is, 141
    what it is not, 142
Finnegans Wake, 136
first law of motion, 126
FitzGerald, Edward
    Rubaiyat of Omar Khayyam, 138
force
    electroweak, 177
Fourier, Joseph, 7, 132
frequency of mind vibration, 1
Freud, Sigmund, 23
Fundamental Fysiks Group, 80, 84
    Capra, Fritjof
        Tao of physics, 80
    Herbert, Nick
        typewriter and spirits, 80
    physics and theosophy, 81
    Sarfatti, Jack
        UFOs and Bigfoot, 80
    Wolf, Fred Alan
        yoga of time travel, 81

G

Galileo, 20, 47, 87, 111, 112, 132, 133, 180
  first law of motion, 119
  and law of inertia, 131
  math as language of Nature, 131, 160
  and reductionism, 119, 126, 151
  what is motion?, 119
Galois, Évariste
  invents group theory
    at 19, 132
Gates, Bill, xii
gauge bosons, 141
gauge fields, 142
  massive
    Higgs mechanism, 144
  massless
    long-ranged force, 143
gauge theory, 142–143, 150
Gell-Mann, Murray, 66, 134
  color charge, 149
  quarks, 136
general relativity
  field equation
    Einstein, 155
    Hilbert, 155
genetic program, 120
Geneva, 148
German Physical Society, 114
Gilbert, William
  De Magnete, 138
Gilgamesh, 11
Glashow, Sheldon, 145
gluon
  and confinement, 149
  gauge bosons
    strong nuclear force, 149
God particle, 148
God plays dice, 44
Goldstone bosons, 144, 148
Goldstone, Jeffrey, 144
Golgi, Camillo, 122
good shepherd, 10

GPS
  and general relativity, 151
GR, *see* general relativity
gravitational waves, 117
Gross, David, 150
  and Geoffrey Chew, 94
group theory, 132
GTR, *see* general theory of relativity

H

Hadith, xiii
hadron, 92, 133, 141, 146, 149
Hahnemann, Samuel, 163
Hamlet
  only one, 156
Hare Krishna, 79
Harkin, Sen. Tom
  bee pollen, 5
Harry Houdini
  and spiritualists, 20
Harvard University
  and alternative medicine, 6
heavy leptons, 146
heavy water, 192
Heisenberg, Werner, 7, 24, 30, 33, 53, 63, 65, 69, 108, 115, 156, 158–160, 182, 194
  Einstein on, 32
  Indian philosophy, 25
  matrix mechanics, 158
  and mysticism, 31
  Post-Materialist Science
    manifesto, 75
  separating faith and science, 26
  1927 Solvay Conference, 25
  uncertainty principle, 31, 52
    Bohr's philosophization of, 54
    for bullets, 53
    for electrons, 53
    turned into complementarity, 61
Hellenistic Greece, 4
Heraclitus, 14

hidden variables
  and Bell's inequality, 46
  and EPR paradox, 46
Higgs boson, 147–149
  discovered, 148
Higgs mechanism, 144, 148
Higgs, Peter, 144
Hilbert, David, 155
Hinduism, 2, 18, 19, 24, 25, 47, 70, 86, 94, 96
holism, 22, 93, 121, 125–127, 150, 193
  attack on reductionism, 129
  basic building blocks
    opposed to, 151
  blocking science, 16
  impedes science, 125
homeopathy, 5, 163
  harm of, 4
  NCCIH, 6
Huffington Post
  Maharishi Years, 98
HuffPost, 117
Hurricane Dorian
  stopped with mind, 2
hydrogen atom
  a gift of Nature, 50
  quantized energies, 50
  quantized radii, 51
hydroxychloroquine, xi

I

IAS, *see* Institute for Advanced Study
Inhofe, Sen. James
  disproves global warming, 6
Inquisition
  chambers, 4
  courts, 4
Institute for Advanced Study, 66
insurrection
  January 6, 2021, xii
intelligence
  quantification of, 122

Intelligent Design
  and Natural Theology, 187
  and scientific creationism, 187
  and Templeton Foundation, 67
interaction
  electromagnetic, 143
  electroweak
    Lie group doublet, 149
  fundamental, 140
  strong, 140
  weak nuclear, 149
interference
  constructive, 51
  destructive, 51
  pattern, 51
    as probability distribution, 59
invisible hand, 39
It from Bit, 71, 74

J

J particle, 146
jinn, 15
John of God, xii
Joyce, James, 136
jumping consciousness, 102

K

Kant, 14, 23
Kepler, Johannes, 132
Kirchhoff, Gustav, 112
koan, 90, 116
  none in physics, 90

L

Laplace, 20, 159
Large Hadron Collider, 130, 148
laser precision
  an oxymoron, 35
Lavoisier, Antoine, 156

Law of Addition of Velocities, 189
law of attraction, 1
Lawrence, Ernest, 129
length contraction, 189
Lenin, 23
lepton, 133, 140–141
    electron, 140
    muon, 140
    neutrino, 140
LHC, *see* Large Hadron Collider
Library of Alexandria, 4
Lie group, 132, 145, 149
    and symmetry, 141
    eightfold way
        particle classification, 134
    for weak nuclear force, 145
Lie, Sophus, 132
life
    what is it?, 120–121
life force, 15, 16, 39
Lord of Dancers, 81
Lorentz, Hendrik, 155
Lorentz transformation, 155

M

magnetic power, 1
Maharishi Mahesh Yogi, 2, 97–99
    death and resurrection, 98
"mainstream" science, 111–112
mana, 15
Manhattan Project, 7, 65, 157
Marx, Karl, 23
mathematics
    language of Nature, 131–134
matter field, 147
Maxwell, James Clerk, 138
    predicts EMWs, 49
    unification of electricity and
        magnetism, 87
Mayo Clinic
    and alternative medicine, 6
Mécanique Céleste, 20, 159

meditation, xi–xiii, 1–3, 80, 116
Mendeleev, Dmitri, 173
meson, 134
miasms, 164
Michelangelo, 155
military-industrial complex, 79
modern gurus
    and quantum physics, 6
modern physics
    as "extra"sensory, 45
modern superstition, 3
Mona Lisa
    only one, 156
monotheism, 15, 16
Moon, 169
    diameter of, 126
    distance, 126
Moses, 2
motion
    what is it?, 119
Mount Olympus, 12
muon
    neutrino of, 140
    second lepton, 140

N

nanotechnology, 45
Napoleon, 7, 20, 159
National Institutes of Health
    and alternative medicine, 6
natural philosophy, 180
Natural Theology
    and Intelligent Design, 187
NCCIH, *see* National Center for
    Complemetary and Integrative
    Health, 5
Ne'eman, Yuval, 134, 172
neo-Nazism, 4
neurotransmitter, 102
neutrino
    detected, 147
    third lepton, 140

New Age, xi–xiv, 2–4, 6–8, 16, 22, 33, 40, 47, 63, 72, 76, 77, 80, 88, 90, 91, 93, 100, 103, 106, 110, 119, 121, 123, 131, 150, 151, 162, 191
   absurd syllogism, 22
Newton, 16, 20, 24, 25, 87, 111, 112, 132, 156, 159, 171, 180
   believing in young Earth, 7
   first differential equation, 132
   and Narendra Modi, 54
   reductionism
      Universal Law of Gravity, 127
Nietzsche, 20, 23
Noah
   copy of Ut-Napishtum, 11
Nobel Prize, 7, 51, 67–69, 71, 74, 90, 93, 94, 107, 122, 145, 146, 150–152, 180, 184, 190
   shared, 156
non-locality, 7, 19, 23, 45–47, 74, 160
   and consciousness, 18
   experimental proof, 46
   lingually inexplicable, 47
nuclear force
   short-ranged, 143
   weak, 149
nuclear fusion
   stellar energy, 128
numina, 15

O

objectivity, 61–63
   definition of, 63
   and determinism, 62
observer effect, 74
Olympia Academy, 23
Ω^-, 135
Oppenheimer, J. Robert, 65
   and Bhagavad Gita, 66
Oprah, ix, xii, 1, 84
   and Zukav, 84

organic, 3
orthomolecular therapy, 7
oxygen
   Lavoisier, 156
   Priestley, 156
   Scheele, 156

P

Paley, William
   Intelligent Design, 187
parapsychology, 2, 85
Parmenides, 14
participatory universe, 103
particle
   classification, 133
Pauli, Wolfgang, 24, 26, 27, 33, 63, 65, 69, 108, 115, 158, 159, 182
   and complementarity, 26
   and Exclusion Principle, 172
   and Jung's synchronicity, 182
   and mysticism, 31, 189
Pauling, Linus
   and orthomolecular therapy, 7
Perl, Martin
   tauon, 146
Persian poetry
   and "third stage", 138
Philosophical Magazine, 157
philosophy, 2, 6, 7, 9, 19, 22, 23, 27, 31
   and mathematics, 181
   atomic, 126
   beginning, 12
   Bohr's interest in, 69
   bootstrap, 151
   claims with no evidence, 115
   Eastern, 83
   founders of quantum physics, 24
   Greek, 12, 123
   holistic, 151
   Marxian, 23
   as opinion, 13–14

natural, 180
Plato, 12, 123
phlogiston, 92
photoelectric effect, 49
photon, 50, 177
   a gauge boson, 141
   consciousness of, 84–86
   is intelligent?, 62
   not a hadron or lepton, 141
   and Yang-Mills theory, 143
physics
   and zoology, 133
   theory
      not "just-a-theory", 149
Planck, ix, 26, 28, 30, 31, 42, 50–53, 108, 112, 113, 158, 159, 161
   Boltzmann, 98
   fighting mysticism, 7
Planck constant, 49–51, 53, 185
   as elemental angular momentum, 50
   reduced, 172
Planck formula
   cosmic microwave background, 113
Planck, Max, 113
   a classical physicist, 87
   and quanta, 49
   praising Boltzmann, 42
Plato, 12–14, 23, 123, 180
   Academy, 13
   disdain for experimentation, 13
   Form, 12
   guardian, 12
   Idea, 12
   and mathematics, 12
   primacy of the mind, 12
polarizer, 59
Politzer, David, 150
pop-spiritualist, xiii, 108, 115
Poseidon, 12
positron
   same as anti-electron, 130
Post-Materialist Science
   a manifesto, 75
potential barrier
   and tunneling, 192
   Coulomb, 128
prayer healing
   researcher's fault, 75
Priestley, Joseph, 156
Princeton University, 66
probability
   alternative medicine, 41
   amplitude, 56
   and life force, 39
   cancer
      survival, 41
   coins, 35–38
   coins to paint, 38
   and consciousness, 39
   dreams coming true, 40
   faith healer, 41
   and invisible hand, 39
   leads to certainty, 39
   quantum, 45, 43–45
   verification of, 63
   verifying its prediction, 36, 183
   weirdness
      macroscopic, 38–39
ψ particle, 146
psi phenomena, 85
psychoanalysis, 23
Ptolemy I
   Alexandrian Museum, 13
Pythagoras, 14

Q

QAnon, xii, 4
QCD, *see* quantum chromodynamics
Qi, 3, 40, 72, 158
   at Harvard, Yale, Mayo Clinic, 6
   NCCIH, 6
quackia
   quanta of fairies, 115
Quackwatch, 5

qualia
  quanta of consciousness, 112
quantum Ayurveda, 101
quantum chromodynamics, 149–150
quantum dots, 184
quantum gravity, 105
Quantum Healing, ix, 2, 27, 97, 98, 100, 158, 190
  and quantum oinking, 100
  dedicated to Maharishi, 98
  Maharishi erased, 98
quantum physics
  "understanding" it, 158–161
  and Upanishads, 18
quark, 92, 136–137, 140
  baryonic charge, 174
  bottom, 147
  charm, 146
  doublet, 146
  down, 136
  electric charge, 174
  Gell-Mann, 156
  model
    and reductionism, 93
  strange, 136
  top, 147
  up, 136
  Zweig, 156

R
Ra, 15, 125
Rayleigh, John W.S., 113
reality, 82–83
  of physics, 83
  of Tao, 82
Red Sea, 2
reduced Planck constant, 172
reductionism, 15, 16, 75, 121, 122, 125, 127, 129, 136, 150, 152, 193
  abhorred by mysticism, 93, 121
  ancient heliocentrism, 126
  essential in science, 125–127
  Greek atomic theory, 126
  leads to atomic nucleus, 127
reiki, 72
Reines, Frederick, 147
relativity
  general
    Einstein, 155
    Hilbert, 155
  a paradox in, 90
  space-time, 118
  special
    Einstein, 155
    Lorentz, 155
    Poincaré, 155
Renaissance, 15, 16, 30, 125, 155
Richter, Burton, 146
Roosevelt, 7, 157
Rutherford, Ernest, 129, 148
  atomic nucleus, 50, 127

S
Saadi Shirazi, 17
Salam, Abdus, 145
Sanskrit
  field theoretic, 95–96
scanning tunneling microscope, 44, 47, 63
Scheele, Wilhelm, 156
Schopenhauer, Arthur, 14, 19, 23, 24, 27, 29, 31, 54, 160, 181
  and acupuncture, 181
  and ESP, 27
  and Jung, 26
  and Pauli, 26
  and Upanishads, 19
  and yoga, 181
  Buddhism and Hinduism, 19
  objectification by Will, 28
  Will and Representation, 19
Schrödinger equation
  epitome of reductionism, 127
Schrödinger equation

and diatomic hydrogen molecule, 128
and helium, 127
and molecular orbital, 128
and valence bond, 128
Schrödinger, Erwin, 24, 28, 63, 65, 69, 108, 115, 156–159, 182, 183
and consciousness, 33, 72
and Deepak Chopra, 27
detached from his equation, 157
and existence of Moon, 28
and Schopenhauer, 28
and Upanishads, 27
his disclaimer, 30
his equation, 23, 44, 47, 185
and astrophysics, 128
and atomic physics, 127
and chemistry, 128
and condensed matter, 128
for double-slit, 56
epitome of reductionism, 127
and H-atom, 127
and lasers, 129
and LED, 129
probability, 44
and reductionism, 122, 129
for solar system?, 47
and transistors, 129
Ψ, 56
science
detached from scientist, 156–158
Science-Based Medicine, 5
scientific method, 131
third stage?, 137–139
scientist
views vs science, 156–157
séance, 20, 23
semiconductors, 129
Shaman, xii, xiii
shamanism, xi–xiii, 81
Sherlock Holmes, 20
Shiva, 81
Dancing, 81–84

SLAC, *see* Stanford Linear Accelerator Center
S-matrix, 93
Socrates, 12
Soviet Union, 23
spectral lines, 51
spin geometry, 158
spirit, 15, 16, 20, 22, 40, 80, 81, 88, 114, 120, 125, 158, 189
as life force, 16
birth of, 14–15
spiritualism, 18, 20, 31
spirituality, 3, 4, 63, 73, 75, 81, 159
New Age, 4
spontaneous remission, 99
spontaneous symmetry breaking, 143–144, 176
spooky action at a distance, 7, 18, 19, 46
SSB, *see* spontaneous symmetry breaking
St. Augustine of Hippo, 123
and Deepak Chopra, 123–124
standard model, 141, 150
and big bang, 150
Stanford Linear Accelerator Center, 146
STM, *see* Scanning Tunneling Microscope
STR, *see* special theory of relativity
strange quark, 136
strangeness, 134
strangeness charge, 134
strong nuclear force
confinement, 149–150
increases with distance, 149
Lie group
triplet, 149
subnuclear physics, 129–131
Sun God, 15
Superconductivity, 122
superposition principle, 56, 57
symmetry, 141
global, 142
local, 142

synchronicity
 "explaining" the irrational, 182

T

Tagore, Rabindranath
 and Heisenberg, 25
Tao of Physics, 2, 81, 93, 150
Taoism, ix, xi, 24, 86, 94
Taoist's denial, 150–151
tau lepton, 146
tauon, 146, 147
Telephone Game, 180
Templeton Foundation
 and Intelligent Design, 67
Templeton Prize, 67
The Secret, 1
theosophy, xiii, 3, 25, 31, 115, 181
 Oppenheimer's influence, 66
thermodynamics
 second law of, 43
thing-in-itself, 14
Thomson, J. J.
 discovers electron, 50
 electron, 107
Thorne, Kip
 and John Wheeler, 71
Tigris, 11
time dilation, 189
Ting, Samuel, 146
top quark, 147
Torah, xiii, 116
transcendental meditation, 2, 97, 99
Trump, Donald, xii, 83
 cancer from windmills, 6
tunneling, 23, 43–45, 47, 128
 lingually inexplicable, 47
 without tunnel, 44

U

uncertainty principle, 24, 52
 lingually inexplicable, 47

universal being, 9
universal field
 of modern gurus, 142
universe
 baryonic part, 106
 composition of, 106
up quark, 136
Upanishads, 18, 19, 27, 28, 54, 181
 and quantum physics, 18
 and Schopenhauer, 19
 Unified Theory of
  and Schrödinger, 27
Υ particle, 147
Ut-Napishtum
 the real Noah, 11

V

vacuum
 in field theory, 144
Vedic sages, 124
 as Einsteins of consciousness, 110
Vedidad-Sade, 17
vibration of the universe, 1
Vietnam War, 2, 79
Vijnana, 15
vitalism, 120
 molecular, 120
von Neumann, John, 65, 66, 69

W

$W^{\pm}$ bosons, 176
 discovered, 145
Wallace, Alfred Russel, 156
warm-blooded plants
 Dyson's idea
  growing on comets, 67
water beetle, 9
water wagtail, 9
Watson, James
 and racism, 7
wave function, 56

and probability, 57
collapse of, 69
  and consciousness, 69
wave-particle duality, 52
weak nuclear force, 145
Weil, Andrew, 5, 112
Weinberg, Steven, 145
Weinberg-Salam-Glashow model, 145
Weyl, Hermann
  admission of guilt, 77
  and consciousness, 70
Wheeler, John, 7, 65, 70, 72, 73, 77, 84, 103, 108, 109, 111, 115, 183, 188
  and Anthropic Principle, 72–73, 105
  and consciousness, 70–72
  and experimenter's consciousness, 74–76
  and general relativity, 71
  the mystic, 71
  and mysticism, 158
  and nuclear fission, 71
  It from Bit, 72, 104, 108
  participator
    being the universe, 103
  participatory universe, 74, 103, 108
  the poet, 71
  Post-Materialist Science
    manifesto, 75
Wien, Wilhelm, 113
Wigner, Eugene, 65
  admission of guilt, 76
  and consciousness, 68–70, 157
  and language of Nature, 133, 135
  and Lie group of relativity, 132
Wilczek, Frank, 150
Williamson, Marianne, 1
Woolley, Leonard, 11
wormhole, 71
WSG, *see* Weinberg-Salam-Glashow
WSG model
  three Goldstone bosons
    four gauge fields, 145
Wu Li Masters, 84–87
  Dancing, 2

Y

Yahweh, 15
Yale University
  and alternative medicine, 6
Yang-Mills theory, 142, 145
yoga, xi–xiii, 2–4
  and time travel, 81
  International Day of, 2
  just a good exercise, 179

Z

Z^0 boson, 177
  discovered, 145
Zarathustra, 20
Zend-Avesta, 17
zoology
  and physics, 133
Zoroaster, 17
Zukav, Gary, 2, 84–87, 189
Zweig, George
  quarks, 136

Printed in the United States
by Baker & Taylor Publisher Services